Information Technology, Innovation System and Trade Regime in Developing Countries

Information Technology, Innovation System and Trade Regime in Developing Countries

India and the ASEAN

K. J. Joseph

First published 2006 by
PALGRAVE MACMILLAN
Houndmills, Basingstoke, Hampshire RG21 6XS and
175 Fifth Avenue, New York, N.Y. 10010
Companies and representatives throughout the world

PALGRAVE MACMILLAN is the global academic imprint of the Palgrave
Macmillan division of St. Martin's Press, LLC and of Palgrave Macmillan Ltd.
Macmillan® is a registered trademark in the United States, United Kingdom
and other countries. Palgrave is a registered trademark in the European
Union and other countries.

ISBN-13: 978–0–230–00492–4 hardback
ISBN-10: 0–230–00492–X hardback

This book is printed on paper suitable for recycling and made from fully
managed and sustained forest sources.

A catalogue record for this book is available from the British Library.

Library of Congress Cataloging-in-Publication Data
Joseph, K. J., 1961–
    Information technology, innovation system and trade regime in
    developing countries:India and the ASEAN/K. J. Joseph.
        p.   cm.
    Includes bibliographical references and index.
    ISBN 0–230–00492–X (cloth)
    1. Information technology—India.   2. Information technology—
    Southeast Asia.   3. Diffusion of innovations—India.   4. Diffusion of
    innovations—Southeast Asia.   I. Title.
    T58.5.J675 2006
    303.48′33091724—dc22   2006043167

10   9   8   7   6   5   4   3   2   1
15 14 13 12 11 10 09 08 07 06

Printed and bound in Great Britain by
Antony Rowe Ltd, Chippenham and Eastbourne

In memory of
my Brother and Father-in-Law

Contents

List of Tables   viii

List of Figures   xi

List of Boxes   xii

Preface   xiii

List of Acronyms and Technical Terms   xx

1   Introduction   1

2   India: An IT Powerhouse of the South   20

3   Thailand: From Investment-led Growth to Innovation-led
    Growth   53

4   Cambodia: Between Pentium and Penicillin?   86

5   Lao PDR: Hastening Slowly?   113

6   Myanmar: Sowing but Not Harvesting?   139

7   Vietnam: Another Tiger in the Making?   169

8   ICT and the Developing Countries: Towards a Way Forward   204

*Appendix   ITA: Addressing Digital Divide Through Trade
Liberalization*   221

Notes   226

Bibliography   238

Index   254

List of Tables

2.1 Export intensity of foreign and local firms in the software
    sector of India (%)   28
2.2 Region-wise distribution of exports by foreign and local
    firms in India (%)   29
2.3 Production and export of electronics products from India
    (US$ million)   32
2.4 Growth of fixed and mobile telephones in India   33
2.5 Network Readiness Index and rank of selected countries
    (2002–2003)   37
2.6 Illustrative S&T infrastructure in four IT clusters in India   45
2.7 Number of foreign collaborations and FDI (approvals) in
    the electrical and electronics and software sectors in India
    during 1991–2002   50
3.1 Structure of exports and imports in Thai electronics
    industry   55
3.2 Stages of latecomer development   57
3.3 Technological stages in Southeast Asia's electronics
    industry   57
3.4 GERD and GERD/GDP of Thailand and other countries
    (2001)   59
3.5 Comparison between the demand and supply of ICT
    manpower in Thailand during 1994–2006   66
3.6 FDI inflows into Thailand in comparison with developing
    countries and Southeast Asia (US$ million)   70
3.7 Sector-wise distribution of foreign investment projects
    approved in Thailand   72
3.8 Distribution of investment across different projects in
    electrical and electronic products in Thailand (2001)   74
3.9 Indicators of telecommunication development in
    Thailand in a comparative perspective (2002)   76
4.1 Higher education enrolments in Cambodia (2001)   93
4.2 Trend in investment approvals and actual inflow in
    Cambodia   97
4.3 Sector-wise distribution of FDI in Cambodia (from 1994
    to 2002)   98
4.4 Cambodia's tariff rate structure   101

4.5	Trend in fixed telephone lines in Cambodia	104
4.6	Distribution of fixed telephone lines across different provinces in Cambodia	105
4.7	Number of telephone subscribers and teledensity in Cambodia	106
4.8	Distribution of mobile market share in Cambodia	107
4.9	Cost of mobile telephone in Cambodia ($)	108
4.10	Internet pricing in Cambodia: an illustration from Camnet	109
5.1	Milestones in the development of telecom sector in Lao PDR	117
5.2	State initiatives towards promoting Internet in Lao PDR	118
5.3	Trends in the approved and actual inflow of FDI to Lao PDR (US$ million)	125
5.4	Inter-regional variations in the access to telecom services in Lao PDR	132
5.5	Trends in mobile cellular penetration in Lao PDR	133
5.6	Indicators of Internet development in Lao PDR	134
6.1	Number of licenses issued for radios, TVs and satellite receivers and the revenue collected in Myanmar	148
6.2	Courses offered at the University of Computer Science	151
6.3	Trend in foreign direct investment in Myanmar: approved and actual (US$ million)	156
6.4	Share of foreign and local investment in different sectors in Myanmar till 2000–2001	157
6.5	Distribution of foreign investment approval across different ownership categories in Myanmar (%)	158
6.6	Trend in export and import in Myanmar	161
6.7	Growth of telecommunications in Myanmar	162
6.8	Inter-regional variation in the teledensity in Myanmar	163
6.9	Cost of mobile telephones in Myanmar (1999)	164
6.10	Number of factories across different size and ownership categories 1998–99	165
7.1	Policy initiatives for the production and use of IT in Vietnam	172
7.2	Key programs and projects identified by the Action Plan in Vietnam	177
7.3	Business cooperation contracts for telecommunication services in vietnam	183
7.4	Trend in FDI commitment and actual inflow in Vietnam (in US$ million)	186

7.5 Relative performance of foreign and local firms in the
     manufacturing sector of Vietnam   188
7.6 Structure and growth of manufacturing output in
     Vietnam (bill dongs at 1994 prices)   192
7.7 Growth of fixed telephone lines in Vietnam   194
7.8 Inter-regional variation in growth of telephones and
     teledensity in Vietnam   196
A.1 Annual average growth rate in exports and imports of
     select ITA products by different group of countries
     (1999–2003)   223
A.2 Number of telephones (fixed and mobile) per 100
     inhabitants in the developing countries of different
     regions in 1992 and 2002   224

List of Figures

2.1 Trend in export and domestic use of software from India   23
2.2 Changing mode of software and service exports   26
2.3 Performance of sectors competing for skilled manpower
    in Karnataka   31
2.4 Teledensity in rural and urban areas of India   34
2.5 Share of IT investment in total fixed investment across
    different industries in India (2001–02)   36
4.1 Growth of fixed vs mobile telephones in Cambodia   107
5.1 Trend in approved foreign and domestic investment in
    Lao PDR   126
7.1 Nominal and effective protection rates by sectors in
    Vietnam: 1997 and 2002   190
A.1 Trend in world export of ITA goods ($ billion)   222

List of Boxes

3.1 Major software initiative: The software park of Thailand 62
3.2 Time frame for project consideration and related procedures in Thailand (Days) 68
6.1 Economic objectives upheld by the Government of Myanmar 140
6.2 Duties and powers of the federation in Myanmar 143
8.1 e-ASEAN Framework Agreement 216

Preface

In February 2003, when I was a Visiting Professor at Jawaharlal Nehru University (JNU), New Delhi, the Trade and Investment Division of UNESCAP approached me to be a consultant. As usual, there was a specific task assigned – undertake a study on "Developing Enabling Policies for Promoting Trade and Investment in the IT Sector of Greater Mekong Subregion" (GMS) and present the results in workshops organized in respective countries and attended by senor/middle-level officers, with local experts acting as discussants, so as to contribute toward policy making in the country concerned. This book is a by-product of that most challenging assignment.

As I have been formulating the specific issues to be explored in the study, I felt that if the study were to deal only with trade and investment policies toward promoting the production and use of ICT in a developing region like GMS, the story is unlikely to be complete. Though the UNESCAP officials agreed to my view, they induced me to focus on trade and investment given the central concern of the division to which I was attached and the time constraint. I was also reminded "there is hardly any study that has looked into all the dimensions of the problem at hand".

This book is an attempt to narrate the "fairly complete" story that I had in mind – the tale of developing a thriving IT production base and promoting its diffusion in developing countries by walking on two legs instead of hopping on one leg and collapsing. At the core of the study is an endeavor to highlight the delicate complementarity between trade and investment regime and the innovation system in promoting the diffusion of ICT without forgoing the returns from domestic production and perpetuating technological dependence. Curiously enough, in the policy parlance these concerns and interrelationships seem to have not received the attention that it deserves. Insights that I gathered through my earlier research on India's ICT sector made me believe that the best way to highlight the above argument is by presenting India's experience in contrast with that of ASEAN countries. Also, there is an imperative of drawing lessons from India's superb performance in ICT software and service sector for other developing countries aspiring to catch up with the ongoing ICT revolution.

As is well known, the period since independence in India saw the building up of a network of R&D institutions, centers of higher learning and other institutional arrangements for building up human capital and promoting the innovation and diffusion process (what is now known as National System of Innovation). Yet, the returns to these efforts remained less remarkable until the trade and investment regime became more liberal in the 1990s. The old ASEAN countries like Thailand, on the other hand, has a longer history of liberal trade and investment regime, at the neglect of building up of National Innovation System, and their achievements in the field of skill-intensive industries like IT and software have not been commendable so far. Hence, these countries are in the process of moving toward the growth strategy wherein innovation takes the prime position. Drawing from the experience of India and Thailand as a setting, the present study undertakes an analysis of the present state of innovation system, trade and investment policies and IT production and use in the ASEAN New Comers to draw lessons for developing countries in general.

Though most of the developing countries have adopted policy measures and initiated institutional interventions to harness ICT as a shortcut to prosperity the outcome has not been identical. In this context, this book argues that a select group of developing countries have acquired substantial capabilities in ICT, unlike the earlier general-purpose technologies, under the Southern Innovation System. Therefore, harnessing the southern innovation system appears to be an unexplored vintage of opportunity to hasten the catching-up process in other developing countries. Whilst the South–South Cooperation has attracted the imagination of policy makers, what is missing today is an appropriate institutional arrangement for exploiting the new avenues opened by Southern Innovation System. Hence, the study makes the case for an e-South Framework Agreement with built-in provision for capacity building along with trade and investment liberalization in such a way as to maintain the delicate balance between trade and investment and building up of an innovation system toward realizing the aspirations of developing countries to harness the new technology for their socio-economic transformation.

Needless to say, writing of this book would not have been possible without the support I received from different institutions and individuals. Though, it is impossible to appropriately highlight the contribution that each of them made, I feel I should at least mention their names without implicating none of them for any of the wrong interpretations, or uncomfortable conclusions, if any, in this book. If a big cheese is missed out, it only shows that my recollection is not that good, as it ought to be. Therefore, I have to begin with an apology to anyone who has been ignored and take full responsibility for anything wrong in this book.

The first string of indebtedness is to UNESCAP and its different officials who liberally spared time to discuss various issues and provided valuable suggestions. Here, I remember with thanks to Dr. Ravi Ratnayake, then Director of Trade and investment division, Mr. Siva Thampi, then Director of Information and Space Technology Division, Mrs. Tiziana Bonapace, Chief, and Dr. Marc Proksch, Economic Affairs Officer and Ms. Marianne Debefve, information Network Officer of Trade and Investment Division. I am especially thankful to Dr. Mathias Bruckner, who dealt with the study on a day-to-day basis and provided very valuable feedback and excellent support.

Though I was associated with JNU during the early days of the study, it was completed and part of the conversion process into the book took place while I was in Research and Information System for Developing Countries (RIS), New Delhi, as Visiting Senior Fellow. Dr. Nagesh Kumar, Director General of RIS, apart from providing me with an excellent working environment, spared many Saturday sessions to discuss "the book project". Other colleagues at RIS, especially Dr. Rajesh Metha, Dr. S. K. Mohanty, Dr. Sachin Chadurvedi, Dr. Ram Upendra Das and Dr. Prabir De, along with their comradeship acted as sources of ideas. Mrs. Sujata Taneja, my secretary at RIS, has shown enormous patience and helped me with some of the routine work involved in preparing a manuscript.

The book was completed after my return to CDS in 2005. Prof. Ashok Parathsarathi, who has been a major source of ideas and encouragement while I was at JNU, acted as a source of inducement in completing the book. In none of our telephonic conversations he forgot to ask, "Joseph, how are you progressing with your new book?" Indeed, my Director, Prof. K. Narayanan Nair, provided me with everything that was needed to accomplish this task at CDS. My colleagues at CDS, especially my *guru* Prof. K. K. Subramanian, Prof. D. Narayana and Dr. K. N. Harilal, not only spared their time for discussions and acted as sources of encouragement but also read through the different parts of the manuscript. Prof. D. Narayana, even read through some parts of the manuscript more than once. Dr. Vinoj Abraham, who was my student in JNU when I initiated the work, became my colleague at the time of completing this book and has been immensely helpful throughout with ideas and support. Mr. Mahesh Sharma, who was also my student at JNU, helped me enormously with different aspects of this work. Mr. Nimesh Chandra and Mr. Manoj Tirkey also helped at different stages. Mr. P. P. Nixon, my Research Assistant at CDS, proved himself really helpful with his hard work.

I have also profited immensely from the intellectual inputs and encouragement received from some of my friends outside CDS. Prof. Amitabh Kundu, Dean, School of Social Sciences, JNU, has been a perpetual source of insights. I will not ever forget his moral support that sustained me at certain hard times. I had very useful discussions with Prof. Hideki Esho, Dean Faculty of Economics, Hosei University. He also shared some important information otherwise not accessible to me. Prof. A. Damodaran, IIM, Bangalore, Prof. Govindan Prayail, Research Director, Centre for Technology, Innovation and Culture, University of Oslo and Dr. Pyarelal Raghavan, currently with Financial Express, patiently read through different parts of the manuscript and offered very constructive comments.

A multi-country study like this would have been impossible without information and insights from different knowledgeable persons from the countries concerned. My fieldwork was coordinated by Mr. I. N. Vothnana, Ministry of Commerce, Cambodia, Mr. Kosei Terami, Program analyst, UNDP, Laos, Ms. Lay lay Su, Staff officer, Ministry of Commerce, Myanmar, and Mrs. Nguyen Thi Thu Houng, Senior Officer, Ministry of Trade, Vietnam, and Mr. Charles Zha, Deputy Director, Foreign Trade and Economic Cooperation, Yunnan Province, and Dr. Matthias Bruckner in Thailand. All of them were at a loss to understand why I asked for more appointments and information beyond the scope of my study. Yet, they cheerfully helped me in getting the needed information. I profusely thank them for their kind cooperation and hospitality and also render my apologies if I had hurt them by putting up an unfulfilled face. In what follows, I shall remember (in no order) the resource persons in each of the country, to whom I am indebted to, for providing me either ideas or information or both.

In Cambodia Mr. Cham Prasidh, Hon'ble Minister of Commerce of the Kingdom of Cambodia, provided very valuable feedback on the chapter on Cambodia. Officials who provided me with much needed information and provided me time for learning from them included Mr. I. A. Alfred, First Secretary, Embassy of India, Cambodia, Dr. Sum Sannisith, Secretary General, National Bank of Cambodia, Mr. Leewood PHU, Secretary General, Ministry of Customs, Economy and Finance, Mr. Uy Sarin, Chief of Customs, Mr. Lar Narath, Under Secretary of State, Ministry of Post and Telecoms, Mr. Chakrya Moa Deputy Director, MPTC, Dr. Hing Thoraxy, Director, CDC, Mr. Chan Sophal

Researcher, CDRI, Mr. Nang Sothy, Deputy Secretary of Executive Board, Cambodia Chamber of Commerce, and Mr. Leng Seyha, Chief of International Relations Department. At the ministry of Commerce I had useful discussions with Mr. Hang Sochivin, Chief, Export Promotion Department, Dr. Pol Neary Tann and Mr. Cham Borith, Deputy Directors at Department of Commerce, and Mr. Yea Bunna, Deputy Director, Industrial affairs Department. Mr. M. Vijaykumar, Managing Director, and Mr. N. Prabhakaran, Manager Finance of Thakral Cambodia Ltd, provided me with great insights into the working of private sector in Cambodia.

Mr. Somphet Khousakoun, Permanent Secretary, Ministry of Foreign Affairs, provided very constructive feedback on the draft chapter on Laos. I also had very useful interactions with Ms. Setsuko Yamasaki, Deputy Resident representative, UNDP, Laos, Dr. Khamlien Pholsena, Director General, Mr. Vilayvong Bouddakham, Director, Department for promotion and management of Domestic and Foreign Investment, Mr. Xaysomphet and Mr. Santisouk phounesavath, Economists, Ministry of Commerce, Foreign Trade Department, Mr. Suksavanh Sayararh, Chief of Trade Information, Mr. Santiphab Phomvihane, Deputy Director General, Customs Dept, Mr. Syliphone Xayavong, Chief, Statistics and Planning, Customs Department, Mr. Soutchay Sisouvong, Deputy Director General, Department of Industry, Mr. Somlouay Kittignavong, Deputy Director General, Department of Science and Technology and Mr. Manohak Rasachack, Officer, Industrial Promotion Division, Industrial Department, Mr. Sumphorn Manodham, President, Wood Products Export Group, Lao Chamber of Commerce and Industry, and Mr. Khamsinthavong Nhouyvanisvong, Secretary General, GMS- BF Secretariat, enlightened me on the working of private sector in Laos. I am especially thankful to Mr. Keonakhone Saysuliane, Director, Department of Science and Technology, and Mrs. Khonesavanh Saysuliane, Managing Director, Infotech computers, for their hospitality and the insights that they provided.

In Myanmar I had discussions with Mr. Nyunt Aye, Director General, Mr. Tint Thwin, Director, Mrs. Thin Htut Thidar, Deputy Director, Directorate of Trade, Ministry of Commerce, Mr. U. Aye Kyaw, Deputy General Manager, Mr. Maung Maung Tin, Managing Director, Mr. U. Tin Tun, Deputy Manager, Mr. U. Tin Myint, Executive Engineer, Myama Post and Telecom, Mr. U. Lin Thaw, Officer on Special Duty, Myanmar Investment Commission, Dr. Maung Maung Lay and Mr. U. Moe, Central Executive Committee Members, Mr. U. Aung Moe, Executive Committee Member, and Mr. Kyaw Nyunt Thein, Executive Officer,

Union of Myanmar Federation of Chamber of Commerce and Industry, Mr. Kyaw U. Thein Oo, President, Myanmar Computer Federation, Mr. Tin Win Aung, General Secretary, Myanmar Computer Association, Thaung Nyunt, General Manager, Myanmar ICT Development Corporation Ltd, Mr. U. Aung Htike, Asst. Research Officer, Department of Advanced Science and Technology, Ms. Daw Kyawt Khin, Professor and Head of Department of Electronics Engineering and Information Technology, Yangon University, Prof. Daw Khin Aye Win, Head, Department of Electrical Power Engineering and Ms. New Ni, Associate Professor, Hardware Technology Department, University of Computer Studies.

In Vietnam, apart from Dr. Le Danh Vinh, Vice Minister of Trade, Government of the Socialist Republic of Vietnam, who provided valuable feedback on the initial draft chapter on Vietnam, I also had very useful interactions with Mr. Sudhir Kumar, first secretary, Embassy of India, Vietnam, Mr. Nguyen Thanh Hung, Director General, Board of Information Technology and E-commence, Dr. Nguyen Quoc Hieu, Chief of Office, Vietnam Electronics Industries Association, Dr. Tran Quang Hung, General Secretary, Vietnam Electronics Industries Association, Mr. Le Hong Ha and Mr. Nguyen Long General Secretaries, Vietnam Association of Information Processing, Mr. Nguyen Van Thao, Deputy Secretary General, VCCI, and Vice President, Vietnam Software Association, Dr. Quach Tuan Ngoc, Director, Center of Information Technology, Ministry of Education, Mr. Nguyen Duc Toan, International Relations Department, Ministry of Education and Training, Prof. Hoang Ngoc Ha, Director of Science and Technology Department, Mr. Nguyen Van Xuan, Deputy Director, Information Technology Department, State Bank of Vietnam, Mr. Richard Jones, Department of Development Economics Research and Consultancy, UNDP, Mr. Lars Heiberg Bestle, Program Officer, UNDP, Mr. Nguyen Thanh Tuyen, and Mr. Nguyen Thanh Tuyen, Officers, Department of Information Technology, Ministry of Post and Telematics, Mr. Lam Quang Minh, Vice Director in charge of Investment Promotion, The People's Committte of Danang City, Mr. Truong Hao, Vice Director, Danang Investment Promotion Center, Mr. Lee Minh Tuan, Head, Project Development Division, Department of Planning and Investment Promotion, Danang Investment Promotion Center, Dr. Ho Si Mau Thuc, Vice Director, and Mr. Ho Tran An, Deputy Manager of Danang Software Park, Mr. Lai Tuan Vuong, Cooperation and Investment Promotion Division, Department of Planning and Investment, HCMC, Mr. Loung Van Ly, Director, Department of Planning and Investment, HCMC, and finally and especially Dr. Tran Ngoc Ca, Deputy Director, NISTPASS.

In Thailand, I have had the benefit of very useful discussion with Dr. Patarapong Intarakummerd, Project Manager of Thailand's National Innovation System Study, NSTDA, Mr. Piromaskdi Laparojkit, Deputy Secretary General, and Mr. Soonthorn Worasak, Director, Board of Investment, Mr. Nadapol Thongmee, Mr. Suwipan Thisyamondol, Directors, and Mr. Rutchaporn Uttranont, Trade officer, Department of Export Promotion, Dr. Arkhom Termpittayapaisith, Senior Advisor in Policy and Plan, NESDB, Dr. Jingai Hanchanlash, Ex. Vice President, Thai Chamber of Commerce, Dr. Kasititorn Pooparadai, Policy Researcher, NECTC.

Finally, the strongest string of indebtedness – the moral support from my beloved wife, Ancy, especially during certain hard times and my loving children, Akhil and Elizabeth. Ancy had to put up with my long absence from home that affected her own work and my children tolerated an absentminded father while at home. My mother, with hardly any understanding of what her son is doing, yet thinking that he is doing something great, is always a source of support.

K. J. Joseph

List of Acronyms and Technical Terms

ACGR	Annual Compound Growth Rate
ADB	Asian Development Bank
ADSL	Asymmetric Digital Subscriber Line
AFTA	ASEAN Free Trade Agreement
AICTE	All India Council of Technical Education
AIS	Advanced Information systems
APEC	Asia Pacific Economic Cooperation
ASEAN	Association of South East Asian Nations
ATM	Asynchronous Transfer Mode
BCCs	Business Cooperation Contracts
BOI	Board of Investment
BPO	Business Process Outsourcing
BTA	Bilateral Trade Agreement
BTO	Build–Transfer–Operate Scheme
CasaCom	Cambodia Samart Communications Company Ltd
CAT	Communication Authority of Thailand
C-DAC	Centre for Development of Advanced Computing
CDC	Council for the Development of Cambodia
CDMA	Code Division Multiple Access
CEPT	Common Effective Preferential Tariff Scheme
CIDA	Canadian Development Agency
CIE	Center for International Economics
CIO	Chief Information Officer
CKD	Completely Knocked Down
CMM	Capability Maturity Model
CSDC	Computer Science Development Council
CSIR	Council of Scientific and Industrial Research
CSO	Central Statistical Organization
DCS	Department of Computer Science
DDFI	Department of Domestic and Foreign Investment
DoE	Department of Electronics
DOI	Digital Opportunity Initiative
ECRC	Electronic Commerce Resource Center
EDC	Entrepreneurship Development Centre

EDI	Entrepreneurship Development Institute
EDPs	Entrepreneurial Development Programs
EHTP	Electronics Hardware Technology Park
EMIS	Educational Management Information Systems
EOUs	Export Oriented Units
EPL	Enterprise of Post Lao
EPTL	Enterprise of Post and Telecommunications Lao
ER&DC	Electronics Research and Development Centre
ETL	Enterprise of Telecommunications Lao
FDI	Foreign Direct Investment
FEA	Faculty of Engineering and Architecture
FECs	Foreign Exchange Certificates
FERA	Foreign Exchange Regulation Act
FIL	Foreign Investment Law
FIMC	Foreign Investment Management Committee
FIPB	Foreign Investment Promotion Board
FOSS-	Free/Open Source Software
FTP	File Transfer Protocol
GDP	Gross Domestic Product
GDPT	General Department of Posts and Telecommunications
GERD	Gross Expenditure on Research and Development
GINeT	Government Information Network
GMS	Greater Mekong Subregion
GPT	General Purpose Technology
GSM	Global System for Mobile Communication
GSP	General System of Preferences
IAI	Initiative for ASEAN Integration
ICORC	International Committee for Reconstruction of Cambodia
ICTC	Index of Claimed Technological Competence
ICT	Information Communication Technology
IDRC	International Development Research Center
IISc	Indian Institute of Science
IIT	Indian Institute of Technology
IMF	International Monetary Fund
IML	Information Markup Language
IPL	Investment Promotion Law
IPR	Intellectual Property Right
ISDN	Integrated Services Digital Network
ISIC	International Standard Industrial Classification
ISO	International Organization for Standardization
ISPs	Internet Service Providers

ITA	Information Technology Agreement
ITEC	International Technical Exchange Cooperation
ITES	Information Technology Enabled Services
IT	Information Technology
ITU	International Telecommunication Union
IXPs	Internet Connection Providers
JETRO	Japan External Trade Organization
LAN	Local Area Network
LANIC	Lao National Internet Committee
LAT	Lao Telecom Asia Co. Ltd
LJTTC	Lao-Japan Technical Training Center
LNCCI	Lao National Chamber of Commerce and Industry
LTC	Lao Telecommunications Company
MCPTC	Ministry of Communication, Post and Construction
MFA	Multi Fibre Agreement
MFN	Most Favored Nation
MIC	Myanmar Investment Commission
MIT	Ministry of Information Technology
MNCs	Multinational Corporations
MoEYS	Ministry of Education, Youth and Sport
MoU	Memorandum of Understanding
MPTC	Ministry of Post and Telecommunications
MPT	Ministry of Post and Telematics
MPT	Myanmar Posts and Telecommunications
MRA	Mutual Recognition Arrangements
MRTPA	Monopolies and Restrictive Trade Practice Act
MSSF	M S Swaminathan Foundation
MTEI	Machine Tool and Electrical Industries
NAFTA	North American Free Trade Agreement
NASSCOM	National Association of Software and Service Companies
NCNST	National Center for Natural Science and Technology
NCST	National Centre for Software Technology
NECTEC	National Electronics and Computer Technology Center
NEM	New Economic Mechanism
NGOs	Non Governmental Organizations
NICs	Newly Industrializing Countries
NiDA	National Information Technology Development Authority
NII	National Information Infrastructure
NITC	National Information Technology Committee
NLDS	National Long Distance Service
NRI	Network Readiness Index

NRIs	Non-Resident Indians
NSI	National System of Innovation
NSTDA	National Science and Technology Development Agency
NTITSD	National Taskforce on Information Technology and Software Development
NUL	National University of Laos
OBM	Own Brand Manufacturers
OCBs	Overseas Corporate Bodies
ODM	Own Design Manufacturer
OEM	Original Equipment Manufacturer
OSPs	Operational Service Providers
PIC	Provincial Informatics Center
PIE	Public Internet Exchange
R&D	Research and Development
RBI	Reserve Bank of India
RGC	Royal Government of Cambodia
RMIT	Royal Melbourne Institute of Technology
RTO	Research and Technology Organization
SDPA	Software Development Promotion Agency
SERC	Super-computer Education and Research Centre
SEZs	Special Economic Zones
SIS	Southern Innovation System
SKD	Semi Knocked Down
SMEs	Small and Medium Enterprises
SPT	Software Park of Thailand
STEA	Science Technology and Environment Agency
STP	Software Technology Park
TDMA	Time Division Multiple Access
TNCs	Transnational Corporations
TOT	Telephone Organization Thailand
TRAI	Telecom Regulatory Authority of India
UNTAC	United Nations Transnational Authority in Cambodia
VAIP	Vietnam Association of Information Processing
VCCI	Vietnam Chamber of Commerce and Industry
VNPT	Vietnam Post and Telecommunications Corporation
VSAT	Very Small Aperture Terminals
WAN	Wide Area Network
WSIS	World Summit on Information Society

1
Introduction

I. ICT: Growth engine of the new century

Advancements in Information Communication Technology (ICT) and the related innovations may be considered as the most important contribution by the last century in the field of technology to the present and beyond. Given the generality of purpose and innovational complementarities, ICT qualifies itself as yet another General Purpose Technology (GPT).[1] On comparing ICT with earlier GPTs, David (1990, 1991) found remarkable parallels in terms of their contribution towards augmenting economic growth and human welfare. In general, it has been argued that ICTs are key inputs for competitiveness, economic growth and development. It offers opportunity for global integration while retaining the identity of traditional societies, increases economic and social well-being of the poor and enhances the effectiveness, efficiency and transparency of the public sector, including the delivery of social services (World Bank 2002). Thus viewed, there is hardly any field of human activity wherein ICT could not have its profound influence *inter alia* by revolutionizing the process of information exchange and thereby reducing the transaction cost.

Many economists, however, were not carried away by the euphoria created by the new technology and also by its potential contribution towards productivity enhancement. Early studies in the US (e.g. Oliner and Sichel 1994, Jorgenson and Stiroh 1995, Sichel 1997) have shown that productivity gains have not substantially accelerated despite increased investments in computers and ICT.[2] However, research undertaken in the last few years have provided us with a wealth of accumulated empirical evidence, especially from the developed countries, at the firm, industry and the economy-wide level, indicating that ICT could contribute significantly to productivity and growth (Pohjola 2001, Link and Siegel 2003, UNCTAD 2003, Indjikian and Siegel 2005). Also, there are numerous cases across the world demonstrating that the developing countries can benefit from increased access to ICT, as much as the rich countries, to address various development issues like enhancing competitiveness, empowering people, improving social service provision and poverty alleviation (Digital Opportunity Initiative 2001). No wonder, the UN Task Force on Science, Technology and Innovation (2005) observes that

> ICT differs from other development sectors and technologies... as accelerator, driver, multiplier and innovator, both established ICTs (radio, television video, compact disc) and new ICTs (cell phones, the Internet) are powerful if not indispensable tools in the massive scaling up and interlinkage of development interventions and outcomes inherent in the goals. (pp. 48–49)

Yet, if the data provided by International Telecommunication Union (ITU) is any indication, unprecedented increase in the use of ICT by the developing countries during the last decade notwithstanding, ICT-induced growth and human welfare is confined mostly to the developed countries and to the more advanced regions in some of the developing countries. To elaborate further, during 1992–2002, the share of developing countries in fixed telephones increased from 21 to 45 per cent, mobiles from 12 to 46 per cent, PC users from 10 to 27 per cent and most impressively Internet users from 3 to 34 per cent (UN ICT Taskforce 2005). Though these achievements are by no means less impressive, data on the number of telephone lines in the developing countries of different regions highlights significant inter-regional variation. The telephone density in Sub-Saharan Africa, Pacific and South Asia, for example, is only one-tenth of that in Caribbean. Thus the observed growth in the developing countries during the last decade has been mainly on account of the pent-up demand in certain urban areas of select developing countries leading to significant intra-national digital divide (Joseph 2005a) and thus pausing an additional digital threat to development (Parayil 2003, 2005).

At the same time, it has been argued that unlike earlier technologies, ICTs have certain unique characteristics that provide the opportunity for the developing countries to harness the power of this new technology and leapfrog (Joseph 2002). Therefore, the challenges of attracting new investment in the ICT infrastructure and the promotion of the use and production of ICT in ways appropriate to developing country contexts notwithstanding, there are real opportunities for promoting ICT production and use by involving public and private sector organizations, NGOs and other stakeholders (Mansell 1999). No wonder, the potential of the new technology for increasing productivity, competitiveness and growth on the one hand and creating new employment opportunities, enhancing the quality of life and over all human welfare, on the other has by now been recognized by policy makers all over the world. As a result, almost all the developing countries have undertaken policy measures and institutional interventions towards promoting ICT use and harnessing this new technology for development. These national and sub-national level initiatives have been complemented by initiatives at the multilateral level like the Information Technology Agreement (ITA) of World Trade Organization (WTO)[3] to promote the use of ICT. Since the ICT use in developing countries is constrained by the affordability, ITA aims at addressing this issue by eliminating the tariffs.

II. Issues and the context

While developing countries in general have undertaken policy reforms and institutional interventions to promote the production and use of ICT, the outcome has not been uniform. In South Korea, for example, a vibrant ICT sector, evolved over the years, accounted for more than 30 per cent of the Gross Domestic Product (GDP) growth during 1980–95 (Feong *et al.* 2001). Similarly in Singapore, a country known for its world-class ICT sector, a study by Wong (2001) finds that the estimated returns to IT[4] capital (37.9 per cent) has been more than twice that of non IT capital (14.6 per cent). Despite the liberal trade and investment policies, the development of ICT sector in Thailand has been low and at best moderate in case of Malaysia (Yue and Lim 2002).

Paradoxically, among the developing countries, India is known for her internationally competitive ICT software and service sector. Over the years the ICT software services and Business Process Outsourcing (BPO) have emerged as a major source of export earning, employment generation and enhancing India's credibility in the world market.[5] While scholars have been apprehensive about the focus of Indian firms on the low end of the software value chain (Arora *et al.* 2001, D'Costa 2004), an estimated index of technological capability in a recent study (Joseph and Abraham 2005) indicated an upward mobility of Indian firms. India is also known for a number of innovative ICT applications in e-governance and poverty alleviation (Kaushik and Singh 2004), though her progress in ICT use, so far, has not been as remarkable as in ICT software and service export. Nonetheless, in terms of the Network Readiness Index (NRI), India occupies 37th rank among 82 countries for which the Index has been constructed (Dutta *et al.* 2003). Surprisingly, the per capita income of countries with comparable levels of the Index has been found to be three to four times higher than India.

In this context, this book seeks answers to some of the issues of current policy relevance for developing countries aspiring to catch up with the ongoing ICT revolution. These issues *inter alia* included; what are the factors that configured India's ICT success and how to unravel the forces that contributed to the present state of laggardness of many developing countries that since long followed a liberal trade and investment policy regime? Are there any lessons from India and these developing countries for those developing counties, which have recently opted for trade and investment liberalization as a short cut for developing IT capabilities? Are there any limits to trade liberalization, as envisaged under ITA, to promote the production and use of ICT in developing countries? If in the affirmative, are there any unexplored avenues that could hasten the process of catching up by the developing countries?

In seeking answers to these issues the present study undertakes an analysis of ICT sector with focus on institutional interventions and policy initiatives in India and five Association of South East Asian Nations (ASEAN) countries wherein Thailand represents the old ASEAN member countries having a longer history of liberal trade and investment policies and the other four (Cambodia, Laos, Myanmar and Vietnam) are the ASEAN New Comers[6] that recently resorted to a liberalized policy regime. In a sense the countries studied represent the broad cross section of developing countries and therefore the conclusions and inferences drawn may be of relevance not only to India and ASEAN but also to the developing countries in general.

The selection of countries for detailed exploration is also guided by the recent developments in India–ASEAN relationship. Since the adoption of look East policy by India in the early 1990s, the ASEAN–India partnership has seen a virtual transformation from a sectoral dialogue partnership to a Summit-level interaction within a decade namely 1992–2002. It scaled new heights with the signing of the Framework Agreement on Comprehensive Economic Cooperation at the Bali Summit in 2003 and later with the adoption of "India–ASEAN Vision 2020" to work towards shared prosperity at the Vientiane summit in 2004. While developments in the global trading environment in the last decade provided the fertile soil for regional integration, the leap forward in ASEAN–India partnership may be seen in the context of confluence in the development philosophy of both India and ASEAN in terms of fostering relationship with the rest of the world and India's capabilities in emerging technologies like ICT (RIS 2004).

With the entry of Cambodia, Laos, Myanmar and Vietnam, ASEAN today comprises a heterogeneous group of countries with high levels of economic and digital divide among them (Joseph and Parayil 2005). While the old members of ASEAN have been highly successful in achieving higher growth performance facilitated *inter alia* by greater integration with the rest of the world, the ASEAN New Comers, with the possible exception of Vietnam, exhibit the characteristics of least developed economies. Not only the per capita income of these countries is at a much lower level as compared to ASEAN-6 but their performance in the social sector also needs improvement. Given their lower levels of development, ASEAN vision for 2020 underlines the need for bridging the development gap between the old and the new members of ASEAN. Initiative for ASEAN Integration (IAI) – ASEAN program for narrowing the development gap between the older and the newer members – focuses on Information Technology (IT) as manifested in the e-ASEAN Framework Agreement adopted by the ASEAN leaders in November 2000. More importantly, one of the objectives of the framework agreement on Comprehensive Economic Cooperation with India has been to facilitate the bridging of development divide between the old and the new ASEAN countries (RIS 2004). Thus ICT appears to be India's short cut to ASEAN, and identification of areas for mutually beneficial cooperation could not only help in strengthening India–ASEAN partnership but also set a model for harnessing Southern capabilities for addressing Southern problems.

III. Analytical framework

The new technology, by revolutionizing the process of creation and exchange of information and thereby minimizing the transaction cost, facilitates the transition of an economy to a knowledge-based economy. In such an economy, the contribution of ICT, as a GPT, may be viewed broadly at two levels: (a) on account of the growth of ICT sector (direct benefits) and (b) on account of the ICT diffusion/use and other spillover benefits (indirect benefits). The former refers to the contribution in output, employment and export earning from the production of ICT-related goods and services (Kraemer and Dedrick 2001). The latter refers to ICT-induced development through enhanced productivity, competitiveness, growth and human welfare on account of the use of this technology by the different sectors of the economy and society.[7] To the extent that greater domestic availability of technology acts as a catalyst in the process of diffusion, the direct and indirect benefits, though different, are interrelated. As Ernst (2001) rightly remarked, enhancing the diffusion of ICT, however, does not imply a neglect of ICT production. Both hang together and need each other.

Returns to production of ICT goods

Studies have shown that in the US, wherein the macroeconomic benefits of ICT revolution are already apparent, ICT industries accounted for about 8.3 per cent of the GDP and nearly a third of GDP growth between 1995 and 1999. ICT production also contributed to lower inflation rates since a growing proportion of economic output has been in sectors marked by rapidly falling prices (US Department of Commerce 2000).[8] Unites States is not the only country that benefited from the production of ICT goods. Production of ICT goods has been a major source of economic output, exports and job creation even in developing countries like South Korea, Singapore, Thailand, Malaysia and others. Therefore, it appears that much could be gained by the developing countries by focusing on the production of ICT goods.

However, it has been argued that production of ICT goods need not necessarily be an easy proposition for the developing countries because industrial structure of ICT goods is highly concentrated with high entry barriers. Industry segments like microprocessors are almost closed because standards are set by the leading US-based ICT players like Intel. Most of the segments of ICT industry are highly capital intensive and scale intensive and require specialized skills that only a few countries can hope to achieve (Kraemer and Dedrick 2001). Moreover, early entrants such as Singapore, Hong Kong, South Korea, Taiwan, Ireland and Israel have preempted many of these opportunities to a great extent.

While there is some merit in the above argument, a closer look at the characteristics of ICT industry would reveal that the doors are not that firmly closed for the New Comers. ICT industry is a multi-product industry and the products may be broadly divided into two categories: ICT goods and ICT services.[9] In each of these broad categories there are a large number of products that vary in terms of technological intensity, dynamism, investment and skill requirements (Joseph 1997). This has made possible the segmentation of the industry into separate, yet closely interacting horizontal layers with greater opportunities for outsourcing and thus transforming a vertically integrated industry into horizontally disintegrated but closely interacting market segments. Moreover, as argued by Ernst (2002) under global production network that characterizes ICT production today, geographical dispersion becomes more concentrated in case of high-precision design-intensive goods where as in case of lower-end products there is high regional dispersion. Therefore, it is possible that the New Comers could enter profitably into some of these product lines depending on their technological capability, human capital availability and the ability to mobilize capital. What is more, in the near future, the demand for ICT goods and services is likely to increase as the rate of ICT diffusion increases both in the developing and developed countries.

Returns to production of ICT services

The recent developments tend to suggest that the primary producing countries have not benefited from globalization. As United Nations Conference on Trade and Development (UNCTAD) (2004a) rightly points out, the secular decline and instability in world commodity prices and resulting terms of trade losses have reduced the import capacity of many developing countries and contributed to increased poverty and indebtedness. Various studies in the framework of value chain have shown that the value retention by developing country producers of commodities has been decreasing (Kaplinski 2000). This situation is further complicated by the emergence of increasingly concentrated market structures at the international level and stringent standards and requirements in the markets of the developed countries. Hence, along with a renewed impetus to consideration of the "commodity problematique" there is also the need for the developing countries to search for new avenues of income and employment opportunities to survive in the less than friendly international trading environment.

Such an inquiry will lead us to the doors of service sectors that make an effective use of the abundant factor in developing countries, namely labor. Economists have long since noted that the services in general are cheaper in developing countries as compared to the developed countries.[10] Yet, these countries have been unable to take benefit of this cost advantage mainly because the export of most of the services called for the cross-border movement of labor. But the movement of labor, unlike capital, was subjected to series of restrictions. Though the process of globalization, which *inter alia* implied the free movement of products and factors, achieved momentum during the last two decades, there have been hardly any relaxations in the restrictions on labor mobility.

However, the advances in ICT has made possible, to a great extent, the "splintering off" of many of the services from its providers which in turn led to what is often called the outsourcing of services.

No wonder India, with its large pool of skilled manpower, has emerged as a preferred location in the international division of labor in knowledge-intensive industries as well as in BPOs. India is not the only country being benefited from opportunities offered by BPOs. Countries like China, Philippines and others are also emerging as providers of BPO services to the developed countries. Given the fact that BPO services are not very skill intensive and most of the required skill could be acquired in a relatively short span of time, the developing countries need to adopt appropriate policies and institutional arrangements to exploit this growing opportunity and avoid the wasteful competition among themselves.

Returns to ICT use

While there were apprehensions about the return to productivity enhancement on account of ICT use, the evidence from the recent cross-country studies shows that the returns to investments in ICT in terms of productivity and growth are substantial.[11] Pohjola (2001) finds the output elasticity of ICT capital as high as 0.31 for the full sample of 39 countries and 0.23 in the Organisation for Economic Co-operation and Development (OECD) sub sample. Another cross-country study by International Monetary Fund (IMF 2001) also has similar conclusions to offer. Country-specific studies like the one for Singapore (Wong 2001) find that the net return to ICT capital (37.9 per cent) is about two and a half times higher than that for non-ICT capital (14.6 per cent). These studies also show that ICT-induced productivity and growth still remains a phenomenon of developed OECD countries and that the developing countries are yet to catch up.

There are also numerous cases to show that developing countries could benefit from increased access to ICT as much as their counterparts in the developed world to address various development issues like empowering people, improving social service provision and poverty alleviation. In Gambia, for example, ICT is being used to achieve better health outcomes. In Chile, significant results are shown in primary-school education with the use of ICT. In Bangladesh it has led to the creation of direct employment for thousands of local women and men (DOI 2001). A study on the use of ICT in four villages in northern Thailand (as quoted in World Bank 2002, NECTEC 2002) also has similar conclusion to offer. In the Indian context the experience of *Gyan Doot* programme in Madhya Pradesh, Internet Kiosks set up by MSS Foundation in Tamil Nadu and *Bhoomi* Project implemented in Karnataka and various other rural ICT projects stand as a testimony to the positive benefits that ICT could impart to developing countries (India, Planning Commission 2001, Kaushik and Singh 2004).[12] Thus, the developing countries have the potential to benefit not only by engaging themselves in the production but also by promoting the use of new technology in different spheres of human life which in turn could act as a catalyst in the process of much needed socio-economic transformation.

Facilitators of production and use of new technology

Here the key issue is how to enable the developing countries to leapfrog in the field of IT by promoting its production and use? World Bank (2000) underlined the role of the following factors: an educated and skilled population that can create and use knowledge; a dynamic National Information Infrastructure (NII) that consists of telecommunication networks; strategic information systems and the policy and legal frameworks affecting their deployment; an interlinked system of research centers, universities, firms and other organizations that can tap into the growing stock of global knowledge, assimilate and adapt it to local needs and create new knowledge. All these can be grouped into what is now referred to in the literature as an innovation system. In addition, there is the need for an economic and institutional regime that provides incentives for the efficient use of existing knowledge, creation of new knowledge and entrepreneurship wherein the trade and investment policies, by creating a more competitive environment, play a key role. The innovation system and the trade regime, however, reinforce each other and therefore the focus on one with the neglect of other could result in a sub-optimal outcome.

Trade and investment

The virtues of trade liberalization that involves removal of tariff and non-tariff barriers have been well articulated in the literature (Dornbusch 1992, Krueger 1997, Srinivasan and Bhagwati 1999). With the removal of tariff barriers, there will be a corresponding reduction in the price of imported goods. Also the removal of non-tariff barriers could lead to enhanced supply and increased access to imported goods and services. The implications of the reduction in price and increased access may vary from country to country and also between sectors within an economy.

Yet, in case of a developing country the following generalizations may be in order. The decline in domestic prices is likely to make the goods and services more affordable to the consumers leading to increased demand and use. The increased access coupled with reduced price could act as a catalyst in the process of diffusion/use of ICT into other sectors of the economy. If the available empirical evidence is any indication,[13] the increased use/diffusion of ICT could help increasing the efficiency, productivity and competitiveness of the ICT-using sectors. The resultant higher output growth could lead to higher income and employment generation in the domestic economy as a whole. This impact is likely to be strong in the case of less developed countries wherein the affordability, on account of low per capita income and higher price, is a major constraint in promoting ICT use.

Second effect refers to the impact on domestic ICT producing sector on account of increased competition and greater access to needed inputs for production that in turn underscore the link between trade and investment. Increased competition, apart from inducing firms to cut cost of production, leads to the exit of inefficient firms and the absorption of their market share by more efficient ones leading to economies of scale and industry level efficiency.

The link between trade and investment is conditioned by the product characteristics and organization of production. This link is likely to be stronger in assembly-oriented industries as compared to process industries. In an assembly-oriented industry like ICT goods, production essentially involves assembling a number of components and subassemblies based on a design. The production of needed components and subassemblies may be highly skill, capital and/or scale intensive that no country could afford to have the capacity to produce all the needed components and other accessories. Hence there is the need for rationalizing their production across different locations. Perhaps, this is what led to the global production networks (Ernst and Kim 2002) and the international division of labor in ICT production. Thus in the global production network, production of each of the component or subassembly is made across different countries according to their comparative advantage such that the overall cost of production is minimized. This essentially means that the production in any country will call for significant imports and bulk of the output will have to be exported to other countries rather than sold in the domestic market. Hence if the production, and therefore investment, in ICT is to take place in any country the trade regime needs to be the one wherein the free flow of inputs into and outputs out of the economy is ensured. Thus viewed, there is an inexorable link between trade and investment, which is apparently much stronger in ICT as compared to most other industries.

Limits to trade and investment liberalization. While the theoretical case for trade and investment liberalization is elegant, when it comes to the experience of developing countries that resorted to trade liberalization as a short cut to prosperity, we have a mixed picture. Here it may be apt to quote Stiglitz (2002),

> Globalization itself is neither good nor bad. It has the power to do enormous good, and for the countries of East Asia who have embraced globalization under their own terms, at their own pace, it has been an enormous benefit. . . . But in much of the world it has not brought comparable benefits. For many, it seems closer to an unmitigated disaster. (p. 20)

After analyzing the trade reform policies in developing countries Rodrik (1992) convincingly concludes that "trade policy plays a rather asymmetric role in development: an abysmal trade regime can perhaps drive a country into economic ruin; but good trade policy alone cannot make a poor country rich" (p. 103). At its best, trade policy provides an enabling environment for development.

The broad conclusion drawn by Ernst (2001) after a detailed analysis of the electronics industries in Southeast Asia is also not much different. It has been argued that export-oriented production can no longer guarantee sustained growth and welfare improvement. Export-led production also faces serious external limitations from volatile global finance, currency and export markets. Three fundamental weaknesses identified by Ernst – sticky specialization on exportable "commodities", narrow domestic knowledge base leading to limited industrial upgrading and limited backward and forward linkages – are important issues to be reckoned with while resorting to a liberalized trade and investment regime.

In case of ICT production, the link between trade and investment notwithstanding, it has been shown that local capabilities are critical for attracting investment and promoting production. In a context wherein low labor cost is taken for granted, the ability of the developing countries to participate in global production network is governed by their ability to provide certain specialized capabilities that the Transnational Corporations (TNCs) need in order to complement their own core competence (Ernst and Lundvall 2000, Lall 2001). Countries that cannot provide such capabilities are kept out of the circuit of international production network despite their liberal trade regime. Also as argued by Cantwell (1995), Dunning (1996), Makino *et al.* (2002) and Pearce (1999), the Multinational Corporations (MNCs) in the recent years have followed the knowledge-based asset-seeking strategies to reinforce their competitive strengths. More importantly, to get rid of the risk of getting locked up at the low end of the value chain and to facilitate movement along the continuum of Original Equipment Manufacturer (OEM) to Original Design Manufacturer (ODM) and finally to Own Brand Manufacturer (OBM) (Hobday 1994), there is the need for building up an innovation system while resorting to a liberal trade and investment regime. In a similar vein, along with numerous studies, a survey by Saggi (2002) concludes that the absorptive capacity of the host country is crucial for obtaining significant benefits from Foreign Direct Investment (FDI). Without adequate human capital or investment in R&D, spillovers from FDI are infeasible.

When it comes to ICT use, lower prices resulting from trade liberalization need not necessarily promote ICT demand and its diffusion unless the developing countries have the capability to use it. Hence trade liberalization has to be accompanied by capacity building such that needed local content is developed and capabilities are created to make its effective use. This calls for complementing the liberalized trade and FDI policies with appropriate policy measures and institutional interventions with respect to education, R&D and human capital such that learning capabilities are enhanced in all parts of the economy – the central concern of studies on innovation system.

National system of innovation

It is by now recognized that an economy's ability to develop an internationally competitive knowledge and skill-intensive sector like ICT and harnessing this new technology for development in a sustained manner depends, to a great extent, on the National System of Innovation (NSI). While the historical roots to the concept of NSI could be traced back to the work of List (1841), the modern version of this concept was introduced by Lundvall (1985) in a booklet on user–producer interaction and product innovation. Freeman (1987), while analyzing the economic performance of Japan, brought the concept to an international audience. He defined NSI as "the network of institutions in the public and private sectors whose activities and interactions initiate, import, modify and diffuse new technologies" (p. 1). The concept of NSI, as defined by Freeman, highlights the processes and outcomes of innovation. Since then there has been burgeoning literature (Lundvall 1992, Nelson 1993, Freeman 1995, Edquist 1997)[14] focusing on different dimensions of the innovation system. Based on the evolutionary approach to innovation, Nelson and Winter (1977, 1982), Nelson (1981, 1995), Carlsson and Stankiewiez (1995), and Carlsson *et al.* (2002) have advanced the technological systems approach focusing mainly on technologies, their generation, diffusion and utilization. The NSI framework was further enriched by studies on regional systems of innovation (DeBresson 1989, DeBression and Amesse 1991) and sectoral systems of innovation (Breschi and Malerba 1997, Malerba 2002, 2004). Thus the innovation system may be supranational, national, regional or sectoral. These approaches complement rather than exclude each other and selection of the system of innovation should be sectorally or spatially delimited depending on the context and object of study (Edquist 1997).

The NSI framework that goes beyond the narrow confines of product and process innovation is an interdisciplinary approach in an evolutionary perspective with focus on interactive learning and innovation within an economy as key to economic growth and welfare. It emphasizes interdependence and non-linearity wherein institutions play the central role. Such a broad approach of NSI framework, as argued by Lundvall (2002), makes it eminently applicable to wide range of countries and situations. The concept of NSI, however, is based on empirical work in developed countries wherein the concept has been used to describe, analyze and compare relatively strong and diversified systems that deal mainly with discontinuous innovations of Schumpeterian type backed by well-developed institutional and infrastructure support (see Lundvall 2002, Arocena and Sutz 2000). But in case of developing countries, not only the institutional context but also the nature of innovations are significantly at variance with developed countries.

There has been a number of attempts to characterize innovation system in the developing countries (Gu 1999, Arocena and Sutz 2000, Ernst 2002 to list a few, Intarakumnerd *et al.* 2002, Lundvall 2002, Razavi and Maleki 2004). In a careful review of such studies, Razavi and Maleki (2004) find two sets of studies on NSI in developing countries. First group of studies have suggested important revisions in the concept of NSI for its application to developing countries (Edquist 1997, Mathews 2001, Viotti 2001). They prefer to focus on differences in patterns of technological development in industrializing countries with industrial countries and therefore the necessity of new conceptualization and deploying new terminology as an alternative to NSI concept used for rich countries. The Second group of studies have basically taken its applicability to developing countries for granted and preferred to deploy the concept of NSI for both developed and developing countries (Alcorta and Peres 1998, Radosevic 1999). Some others believe in the usefulness of the NSI concept for developing countries and have tried to adopt this concept to the new conditions while, at the same time, clarifying its implications. For example, Arocena and Sutz (2000) point out that industrial innovation in developing countries is highly informal, that is, not the product of formally articulated R&D activities.

The relevance of the broad framework of NSI to countries at different stages of development notwithstanding, each country has its own specific patterns of innovation, technological and trade specialization and specific institutions. Hence, while approaching the issue of catching up one has to search for generalizations within the broad framework. This calls for characterizing the innovation system that is present in those developing countries with some success in catching up so that lessons could be drawn for other developing countries lagging behind in catching up.

Southern innovation system (SIS)?   During 1980s, there has been a growing body of literature that dealt with innovation in the Newly Industrializing Countries (NICs) and their technological capability building process (Cooper 1980, Bell *et al.* 1982, Dahlman and Westphal 1982, Enos 1982, Lall 1982, 1987, Dahlman 1984, Katz 1984, Fransman 1986, Bell and Pavitt 1992, 1997, Patel 1995, Subrahmanian 1995, Kumar and Siddharthan 1997). By analyzing the "inducement mechanisms and focusing devises" along with the process involved and the outcomes of innovation in the NICs, these studies have thrown up significant insights that are highly relevant today in characterizing the process of innovation in the catching-up countries and drawing inference for the less successful countries.

From the literature on technological capability in NICs, it is evident that the inducement for innovation has been the requirement for generating basic production capabilities wherein "know how" rather than "know why" has been the focal point.[15] Yet there are a number of instances in which the developing countries have successfully developed both. Kim (1980), for example, captured a three-stage process of mastery of technology – implementation, assimilation and improvement. In the similar vein, Katz (1984) shows that innovation follows sequence of simple to more difficult activities, namely product design engineering, process engineering and production planning engineering. There are also evidence to show that the government policies acted as a major source of inducement in promoting the innovative process and capability building. Kim (1980) suggests that government support has been essential, especially in the first phase (i.e. imitation) of technological mastery. Other studies also highlighted the importance of government policies by providing a stable and large demand along with easy credit, R&D subsidy, among others, and helping acquisition of technology (Enos 1982, Fransman 1986, Patel 1995, Subrahmanian 1995, Kumar 1998, among others).

The studies on technological capability in NICs also highlighted two different, but interrelated, processes of technological-capability building. The first one related to the transfer of technology through various forms and mechanisms, ranging from FDI and technology licensing which are market mediated with active role to foreign firms to transfer of knowledge through imitation which is non-market mediated and with passive role for the foreign firms from the developed countries (Dahlman and Westpal 1982). Regardless of the mode of transfer, the technology acquired from the developed countries needed several types of adaptations for use in the developing countries. As Nelson and Winter (1977) pointed out, technological knowledge, because of its complexity, cannot be transferred in its entirety. The result is that the purchaser of technology always receives less complete information set than what is possessed by the seller. This forces the technology-importing countries to develop local technological capability through R&D effort. Moreover, since the technology imported has been generated in the developed economy context the effective use of technology transferred called for adaptations to suit the local conditions like the need to scale down technology (Katrak 1985) unavailability of the needed raw materials and spares (Page Jr 1979, Katz 1980, Desai 1984) or the need to diversify the products and to increase the capacity utilization. The second process, therefore, has been the extent of domestic R&D effort, mostly adaptive, by the developing countries.

These adaptive R&D efforts by firms – along with interactions, though limited, with other actors[16] like the universities and the R&D laboratories – resulted in innovations specific to developing countries. These innovations, following Freeman (1986), have been in the nature of incremental improvements in products and processes associated with increasing scale of investment and learning from experience of production and use in contrast to radical innovations in products and process which often were most noticed. Here the underlying processes that generated innovations have been mainly through learning by doing, using and interacting along with STI-mode of innovation (Jensen *et al.* 2005) with emphasis on promoting R&D and creating access to explicit codified knowledge. Though incremental in nature, as a result of their cumulative effect some of the NICs have emerged as technology exporters and there has also been substantial increase in the technological intensity of their exports (Lall 1987, 1992).

The discussion so far dealt mostly with manufacturing technologies and of late there are a number of instances wherein the developing countries have built up substantial capabilities in the new and emerging technologies like ICT, which could also be of immense importance to countries less successful in catching up. In a sense, the potential of the new technology to contribute to the socio-economic transformation of the developing world emanate from the fact that while the developed countries held monopoly over the earlier GPTs, in case of ICT, substantial capabilities are present in the South (Joseph and Intarakumnerd 2004). While Japan and Southeast Asian countries hold leading position in the manufacture of ICT goods (Ernst 1993) and China has recently joined the league, in the field of ICT software and services India has emerged as a major player in the world market.

India is not an isolated success story in the South. A number of non-G7 countries have developed capabilities in the field of ICT and software (Arora and Gambardella 2004) and a new generation of countries like Philippines, Morocco, Costa Rica and others are joining the bandwagon (UNCTAD 2003). Capabilities in ICT that emerged from the South are manifested in hardware innovations like Simputer, CorDECT Wirless in Local Loop and various innovations in the field of free and open source software. Interestingly, these innovations address the issues specific to developing countries like affordability, illiteracy and connectivity (Joseph 2005b). Thus an understanding of the innovation system that facilitated the capability building and its diffusion may be of immense relevance for countries that have been less successful in catching up. This becomes all the more relevant because, as argued by Koanantakol (2002) the elements and priorities of national ICT strategies of developed countries (IT security, privacy, cross-border certification, etc.) are at variance with that of developing countries like basic telecommunication, affordability, local content, human capital and others.

Drawing from the above discussion, it may be inferred that the broad concept of NSI is eminently useful in understanding innovation in developing countries as much as in the developed. At the same time, given the different inducement mechanisms along with the processes and outcomes in developing countries and their bearing on informed policy making, there is the need for articulating the concept of SIS and this is yet to receive the attention that it deserves. The broad contours of SIS involve, but not limited to, its predominance in catching-up countries induced by the necessity to develop basic production capabilities, and SIS has been an outcome of the interaction between technology import and adaptive R&D in manufacturing industries and later spilled over to service sectors like ICT. The outcome has been incremental in contrast to radical innovations, and their cumulative effect has been manifested in the export of technology on the one hand and increased technological intensity of their exports on the other. The whole process has been less market driven, as the market mechanisms in developing countries have been underdeveloped and government policies playing an important role.

If the available evidence is any indication, the outcomes of SIS are increasingly being harnessed by the MNCs through mergers and acquisitions and setting up R&D Centers in catching-up countries like India, China and others. In the computer and telecommunications sector, foreign investors have established over 200 R&D centers, programs, or labs in China between 1990 and 2002. More than half (128) of these were established by US multinationals (Walsh 2003). In the past five years, more than 110 Multinational Enterprises (MNEs) have set up R&D centers in India including the GE Capital's $80 million technology center at Bangalore, that is the largest outside the US, employing 1600 people.[17] Indian R&D centers of the US MNEs have filed more than 1000 patent applications with the US PTO mostly during 2002–2003 (Kumar and Joseph 2005).

Thus viewed, the developing countries of today are endowed differently as compared to their counterparts who achieved political independence after the Second World War. Fifty years back the developing countries' choice with respect to sources of technology was limited to the technology shelf of developed countries. In contrast, today there are more options open to countries that are less successful in catching up. As we have already seen, during the last 50 years a number of countries have developed substantial technological capability that evolved under the SIS. Evidently, these capabilities could be readily applicable to countries that are less successful in catching up. In addition, these developing countries could also choose from the technology market of developed countries and, as argued by Ernst (2002) could benefit from the new opportunities offered by global production network. The extent of success that the developing countries are likely to achieve with respect to the promotion of production and use of ICT on the one hand and the catching-up process on the other would depend to a great extent on how intelligently they make use of these alternate avenues open to them.

IV. Reader's guide

The present study is presented in eight chapters including this introduction and the final chapter wherein a perspective for the developing countries is presented drawing from the experience of India and the ASEAN countries. The second chapter presents the growth performance of ICT sector in India with focus on software and services and highlights the contributory factors using the framework of NSI. The experience of Thailand, which adopted a strategy different from India with respect to ICT production and use, forms the focus of the third chapter. Against this background, chapters four to seven present the case studies of Cambodia, Laos, Myanmar and Vietnam respectively.

In the case studies of ASEAN New Comers, answers to the following three broad questions are sought. First, what is the present state of innovation system in the ICT sector? Secondly, to what extent the trade and invest regime in the country is conducive to develop an innovation system that would promote the production and use of ICT in different sectors of the economy? Finally, what is the extent of success that these countries have achieved with respect to ICT production and use? To answer the first issue, a detailed analysis of (a) different policy reforms and institutional interventions, (b) present state information infrastructure and (c) initiatives and achievements with respect to human resource development for the ICT sector has been undertaken. The second issue has been addressed by a critical analysis of the trade and investment policies and their bearing on building an innovation system in the ICT sector and their success in attracting investment and promoting trade in ICT. An analysis of the present state of ICT production and trends and patterns in the use of telephones (both fixed and mobile), computers and Internet has been undertaken to seek an answer to the third issue.

The last chapter, by drawing inference from the case studies and a critical analysis of the outcomes of Information Technology Agreement (ITA) of WTO, highlights the limits to trade liberalization as a strategy to promote the production and use of ICT and calls for complementing these measures with initiatives to build up the NSI. In this process the study underscores the need to harness the Southern Innovation System (SIS) to hasten the process of catching up by the developing countries in ICT. But what is currently missing is an institutional arrangement making effective use of SIS. Given the limits to ITA under WTO, and regional cooperative arrangements like e-ASEAN Framework Agreement the study makes the case for an e-South Framework Agreement among countries to promote ICT use and production through trade liberalization and capacity building such that the digital divide is transformed into digital dividend.

2
India: An IT Powerhouse of the South

I. Introduction

There is an often-held view that India undertook pro market reforms in the 1990s and this has led to a revival of the Indian economy in general and the emergence of some of the star-performing sectors like IT software and services that contributed to a change in India's image in the rest of the world. But, for a careful observer, reforms have had their beginning in the 1980s and as a result of these reforms coupled with the sound technological, industrial and human capital base built up over the years the Indian economy began to show signs of turnaround as early as in the 1980s. As Rodrik (2004) rightly states, India, despite the folk wisdom that relates its growth acceleration to the liberalization of 1991, actually began its take off a decade earlier.

After recording an annual growth rate of 2.98 per cent during 1970–80, the Indian economy revived during the 6th (1980–85) and 7th (1985–90) plans recording an annual growth rate of 5.5 and 5.8 per cent respectively. The initial two years of 1990s turned out to be the difficult years mainly on account of the crisis in the external sector and there has been a drastic decline in the growth rate. During the 8th plan (1992–97), subsequent to the major policy initiatives as part of the stabilization-cum-structural adjustment, the economy recorded a marked revival with a growth rate of 6.8 per cent. The higher growth rate, however, could not be sustained during the 9th plan (1997–2002) mainly on account of the lower growth of the world economy along with poor agriculture resulting from adverse weather conditions and to a lesser extent due to East Asian financial crisis. However, in 2003–2004 and

2004–2005 the economy surged ahead recording an annual growth rate of over 8 per cent. In general, during 1980–2004 the Indian economy recorded an annual growth rate of 5.9 per cent (Joseph 2004d).[1] Such a high growth rate sustained for more than two decades appears to be unprecedented in the Indian economic history and second only to China. At the same time, India's growth during the last two decades was shown to be most stable, surpassing even China and the East Asian Countries (Rodrik and Subramanian 2004a).

There has also been major structural transformation of the economy in terms of the distribution of GDP across different sectors of the economy. In 1980 agriculture was the largest sector of the Indian economy contributing about 40 per cent of the GDP followed by services (36 per cent) and the industrial sector (24 per cent). By 2002 the share of agriculture in GDP declined substantially to reach a level of 23 per cent. While the share of industrial sector remained more or less constant, that of service sector increased substantially to contribute over 56 per cent of the GDP. Thus the growth of Indian economy in the recent past has been driven by service sector and it is known world over for its remarkable performance in emerging service sector like software and BPO.

Perhaps, the most impressive aspect of India's economic performance during the last decade related to the external sector. From a situation of extreme crisis in 1991 with foreign exchange reserves hardly sufficient to meet four weeks' imports and the current account deficit as high over 3 per cent of GDP, there has been a remarkable turnaround with surplus in current account and the foreign exchange reserves exceeding $140 billion in 2005. Very few examples exist, if at all, in the developing world of such a turnaround in a short span of time.

Driven by the abundant supply of skilled manpower at competitive rates[2] and other institutional infrastructure built up over the years, India emerged as a major player in some of the skill-intensive industries like automobiles and parts, drugs and pharmaceuticals and others. Moreover, being a country that produces one out of every six engineers in the world, India has become the world's R&D hub wherein MNEs, some of them with R&D budgets larger than India's total R&D budget, have been setting up global or home-base augmenting R&D centers. In the past five years, nearly 110 MNEs have set up R&D centers in India. These include GE's $80 million technology center in Bangalore, which is the largest outside the US and employs 1600 people. Indian R&D centers of the US-based MNEs have filed more than 1000 patent applications with the US Patent Office mostly during 2002 and 2003. The Indian centers of multinational technology companies expect to double the number of their employees in the near future from 40,000 in 2003. According to a survey by McKinsey, of 5500 corporate leaders with billion dollar revenues, 31 per cent of executives favored R&D investments in India compared to 27 per cent in China.[3] India's private sector has been acquiring Western firms and India has emerged as one of the leading developing countries with outward investment (Kumar and Pradhan 2003).

No wonder, the analysts are competing among themselves to present optimistic predictions on India's future growth. Wilson and Purushothaman (2003), basing primarily on favorable demographics of the country, forecast a growth rate of 5–5.5 per cent per annum over the next 30 and 50 years. According to them, India's GDP is likely to exceed that of Italy in 2020, France in 2025, Germany in 2030 and Japan in 2035! The road ahead is less rocky and destination too close according to Rodrik and Subramanian (2004b). Using a simple growth-accounting framework, they predicted a growth rate of 7 per cent per year for output and 5.6 per cent for per capita output for the next 20 years.

While there are number of sectors wherein India could claim notable achievements during the last two decades, performance in IT and more specifically in IT software and service sector has undoubtedly been the most noted. Hence it is only natural for other developing countries to look towards India for lessons for possible emulation. In this context, the second section of this chapter undertakes an analysis of the broad contours of India's performance in the ICT sector covering both software and hardware sectors and examines the present level of ICT use. Studies have highlighted a number of factors that *inter alia* include the availability of highly skilled manpower at the fraction of their cost in developed countries, proficiency in English language and links with diaspora as factors that contributed towards India's performance (Singh 2002). While there is merit in these arguments, it has also been claimed that India's performance has been an outcome of benign state neglect and greater role of market forces. Given the fact that there are many developing countries that adopted liberal trade and investment regime but with limited success in developing internationally competitive skill-intensive sectors like ICT software and services, the third section makes an attempt to exploring the role of national innovation system and the trade and investment regime in shaping India's ICT success. The last section presents the concluding observations.

II. ICT production, export and use

ICT software and the economy

India's IT software and service sector, which includes the delivery of BPO, has grown at an unprecedented rate over the past decade or so. The value of output of India's software and service sector increased by more than 18 times from less than $0.83 billion in 1994–95 to $15.5 billion in 2003–2004. That the growth of the sector was fuelled mainly by the exports is clear from the fact that exports of software and services increased by more than 25 times during the same period (Figure 2.1). Such a high export growth, sustained for more than a decade, has been unprecedented in India not only in terms of the magnitude of the observed growth rate but also in terms of its stability. With such a higher growth rate, the software and service sector accounted for over 2.6 per cent of GDP, in 2003–2004 compared to 0.5 per cent in 1996–97,

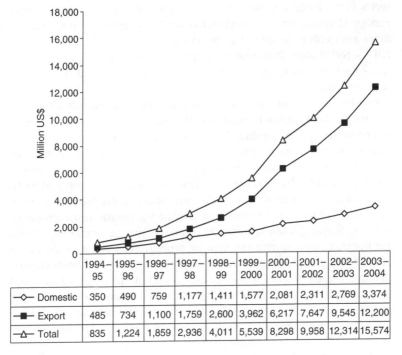

	1994–95	1995–96	1996–97	1997–98	1998–99	1999–2000	2000–2001	2001–2002	2002–2003	2003–2004
—◇— Domestic	350	490	759	1,177	1,411	1,577	2,081	2,311	2,769	3,374
—■— Export	485	734	1,100	1,759	2,600	3,962	6,217	7,647	9,545	12,200
—△— Total	835	1,224	1,859	2,936	4,011	5,539	8,298	9,958	12,314	15,574

Figure 2.1　Trend in export and domestic use of software from India
Source: NASSCOM (2004).

representing an increase over five times, and generated about 0.6 million employment in 2003–2004. Its share in India's exports has increased almost sevenfolds (from 3.2 to 21.3 per cent) during the same period and has contributed significantly towards the turnaround in India's current account.

More importantly, India established credibility in the international software and service market, which in turn has induced over 300 *Fortune 500* companies to outsource software services from India. The credibility has been established on account of the better quality services and there is hardly any IT firm in India that does not hold necessary quality certification. What is more, as many as 50 Indian companies have received SEI-CMM (Software Engineering Institute, Capability Maturity Model) Certification at Level 5 out of 74 such organizations worldwide or out of 54 outside the US (NASSCOM 2004b). This shows that Indian software enterprises, especially the leading ones, have strived to attain excellence in their professionalism and world-best practices. Indian IT firms have also been acquiring foreign firms in large numbers and the IT sector is the leading sector with outward investment from India. India exports IT and IT-enabled services to over 133 countries and the Indian firms are training people in IT in 55 countries. A single Indian firm – NIIT – today runs 100 training centers in China alone! The government itself is setting up training centers for people in other countries (e.g. Mangolia).

Besides creation of jobs for highly qualified professionals as well as ordinary college pass-outs, the rise of the software industry has provided opportunities for expanding the local base of entrepreneurship. The initial start-up costs in the sector are rather low and economies of scale are not particularly significant, especially for service enterprises. Hence, the entry barriers are low. This has helped a number of technical professionals to start on their own. Many of the leading software enterprises of today have been started by first-generation entrepreneurs. Infosys, Satyam, Mastek, Silverline, Polaris, among numerous others, for instance, were started by software professionals and engineers with small savings and loans at very modest scales to begin with (Kumar 2001a). A study of smaller or informal sector enterprises in software and services industry in India corroborated the rewarding opportunities for entrepreneurship for little initial set-up costs. The rates at which even these smaller enterprises have been growing means that they do not stay small for very long (Kumar 2000).

The rapid rise of the software industry in the country has also helped to reduce the extent of brain drain by creating rewarding employment opportunities within the country, a trend also supported by the availability of venture capital to implement new ideas. It has also prompted a number of Non-Resident Indians (NRIs) to return to the country to start software ventures. Apparently, in Hyderabad alone about 100 companies have been set up by returning software professionals (Kumar 2001a). Furthermore, the export-orientation of the Indian software industry benefited from the presence of a substantial number of NRI engineers working in US MNEs. Arora *et al.* (2001) observe that some of them have played an important, though yet to be documented, role in facilitating the contacts between buyers in the US and the potential suppliers in India. NRIs in the software industry in the US have also invested back home in subsidiaries that develop software for their US operations. These include investments in subsidiaries of Mastech, CBS Inc., IMR, among others.

From software to IT Enabled Services (ITES)

India, with 253 universities and 13,150 colleges, produces about 2.46 million graduates and about 290,000 engineering degree and diploma holders every year. English is widely used as a medium of instruction, which in turn provides an ample supply of manpower for ITES at a much lower cost as compared to other countries.[4] Hence a new development that is taking place in the industry has been the emergence of BPO through Internet or the so-called ITES as a more dynamic component of Software and service exports from the country. In 1999–2000 the total ITES exports was only of the order of $565 million and it is estimated to have increased to $3.6 billion in 2003–2004. ITES exports grew by 59 per cent during 2002/2003 to 2003/2004. Hence, but for the rise of BPO exports, the growth of software exports would be even lower than what is shown above. As a result, the share of ITES in total software and service exports more than doubled from about 14 per cent in 2000 to more than 29 per cent in 2003–2004.

Though the ITES/BPO services, experiencing a boom at present, are relatively low value-adding and low skill-intensive activities, they have certain characteristics that could contribute to broad-based development. While employment in the software sector has been mainly for the highly skilled IT professionals, the ITES sector generates more broad-based employment including the arts and science graduates. It is also found that ITES sector is more employment intensive with employment per million dollars of exports as high as 70, which is more than twice that of the software sector (Joseph 2004a). Thus viewed, ITES/BPO appears to have the potential of generating substantial employment for the growing number of educated youth in the country. Further, ITES calls for relatively less institutional linkages as compared to software. Hence, for those regions, which were not successful in attracting software investment, ITES offer an alternative. Given its specific characteristics and policy initiatives by the regional governments the ITES activity is likely to be more diffused across different regions in the country and generating more linkages with rest of the economy.

Moving up the value chain

In the early years of its development, the software and service exports from the country were carried out mostly in the form of onsite development (Heeks 1996). However, with the setting up of a number of Software Technology Parks (STP), which *inter alia* provided access to modern telecommunication facilities, and liberalized policies towards the telecom sector, which in turn led to the entry of a number of private sector telecom companies, there has been a significant shift away from onsite development. Today nearly 60 per cent of the exports takes the form of offshore development (Figure 2.2).

So far, most of the Indian software enterprises have been focusing on services that are considered to be low value adding. Having got

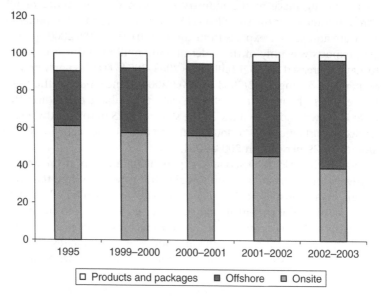

Figure 2.2 Changing mode of software and service exports

themselves established as suppliers of these services, Indian companies are now making a conscious effort to increase exports of high-end consulting with the development of domain expertise and export of packaged software. Infosys, for instance, is focusing on the export of end-to-end services. As Indian software enterprises establish their credentials and competence, they are consciously seeking fair value for their work. This, however, may be applicable for leading companies such as TCS, Infosys, HCL Technologies, WIPRO, Satyam Computer Services, which are providing higher-end programming solutions to their clients. Infosys has successfully renegotiated its per man-hour charges with its clients. It reportedly commands $90 per hour (Kumar and Joseph 2005).

Indian companies have also managed to develop a number of proprietary software products. A niche market has been created in banking, financial and accounting software. These include, for instance, I-Flex that has been used by over 240 financial institutions in 69 countries. Polaris developed a proprietary retail banking software, Polaris Point, and is tying up with Bull, France, for its marketing in Europe. Banking solutions from Infosys (Financle, Bankaway and Payaway) have been adopted by 22 domestic and 16 overseas banks across 12 countries as early as in 2001. TCS launched packaged software for banking insurance, securities, accounting and health care industries. TCS also launched its branded integrated suite of software tools, Mastercraft, which is claimed to have been received well in the US and Europe and carries a price tag of US$150,000. WIPRO Technologies launched a number of branded products including Teleprodigy, a billing system for Internet Service Providers (ISPs) and WebSecure, an Internet security package. It is focusing on global brand building and plans to come up with a branded product every year. A number of even smaller software companies have developed packaged software, which is sold in domestic market. For example, Tally, a popular accounting package for SMEs which is being used by 50,000 companies and has been approved by the Accountants' professional bodies in India and the UK, has been developed by a smaller highly specialized software company (Kumar 2001a).[5] Despite these efforts, as evident from Figure 2.2, share of products and packages in the Indian exports of software is still marginal at 3.2 per cent.

Since the conventional measures of innovation like R&D intensity (measured as research expenditure as proportion of sales) has certain limits in capturing innovation in a service sector like ICT, a study (Joseph and Abraham 2005) developed an Index of Claimed Technological Competence (ICTC) using firm-level information on their areas of specialization. The theoretical base of the index has been drawn from the literature on technological opportunity. The estimated index has shown that firms are moving up the value chain.

Locally anchored capability

The highly facilitating policy towards FDI along with an abundant supply of skilled manpower has induced a large number of foreign firms to enter India's IT and software service sector. The first foreign firm to enter India's software sector is Texas Instruments (TI) in 1984. TI was followed by other IT and software majors in the world and today there is hardly any leading IT firm in the world that does not have presence in India. The presence of foreign firms in India naturally raised India's creditability in the world market and would have led to spillover benefits. Over the years, the role of FDI increased in India's software sector and it has been claimed that in states like Karnataka[6] and Maharashtra, three foreign companies are entering the software and service sector every fortnight.

An analysis of the firm-level data compiled from National Association of Software and Service Companies (NASSCOM) directories reveals that foreign firms are more export oriented as compared to the locals firms. The export intensity (measured as a proportion of output being exported) is found to be as high as 92 per cent for the foreign firms whereas the corresponding share for the local firms is about 70 per cent, and of late there has been an increase in the domestic market orientation of domestic firms (Table 2.1). The table also shows that the foreign

Table 2.1 Export intensity of foreign and local firms in the software sector of India (%)

Location	Foreign		Local	
	2001–2002	2002–2003	2001–2002	2002–2003
Delhi	97.45	95.11	69.78	72.27
W. Bengal	99.21	98.34	73.67	85.76
Gujarat	0	0	78.34	74.72
Maharashtra	85.45	85.87	74.98	76.38
Andhra Pradesh	98.11	98.44	87.62	87.29
Karnataka	94.56	94.48	76.83	76.63
Tamil Nadu	80.64	89.76	75.32	77.27
Kerala	0	0	80.34	84.62
Others	92.45	80.34	69.23	70.23
Total	92.23	93.88	74.29	70.63

Source: Estimates based on data compiled from NASSCOM Directories.

firms in Maharashtra are more domestic market oriented as compared to their counterparts in other states. This perhaps points towards greater demand forthcoming from the state that is more industrially developed.

Table 2.2 presents data on the region-wise export by foreign and local firms. The table further confirms the high regional concentration of exports. Even within each state, one or two metropolitan centers account for bulk of exports. Thus, Bangalore and to a lesser extent Mysore in Karnataka, Mumbai and Pune in Maharashtra, Noida and Gurgaon near Delhi area, Hyderabad in Andhra Pradesh and Chennai in Tamil Nadu are the centers of software and service activities in the country. The operation of the foreign firms is also confined to these centers.

Software export by the foreign firms during 1997–02 recorded an annual compound growth rate of 51 per cent, which is found to be higher than that recorded by the local firms (30.5 per cent) and by the industry as a whole (33.5 per cent). Thus in terms of the growth rate it appears that during the last five years the foreign firms have been more dynamic as compared to their local counterparts. Even then the foreign firms accounted for only about 20 per cent of the total exports in 2002 (Joseph 2004a). Thus, as argued by Arora and Athreya (2002), India's achievements become all the more striking when noted against the fact that these were made mostly at the instance of local firms.

Table 2.2 Region-wise distribution of exports by foreign and local firms in India (%)

Location	Foreign		Local		Total	
	1997–98	2002–2003	1997–98	2002–2003	1997–98	2002–2003
Delhi	23.39	19.12	11.27	14.68	12.61	15.59
W. Bengal	0.71	0.49	1.38	0.82	1.31	0.74
Gujarat	0.01	0.01	0.17	0.11	0.15	0.03
Maharastra	18.96	27.51	44.40	33.70	41.59	32.43
Andhra Pradesh	1.40	2.68	4.97	9.64	4.57	8.22
Karnataka	33.11	26.99	25.01	31.58	25.91	30.64
Tamil Nadu	19.89	21.68	12.43	9.17	13.25	11.72
Kerala	0.00	0.00	0.16	0.23	0.14	0.19
Others	2.53	1.59	0.21	0.14	0.47	0.44
Total	100.00	100.00	100.00	100.00	100.00	100.00

Source: Estimates based on data compiled from NASSCOM Directories.

Opportunity cost of exports and domestic linkages

The opportunity cost of software exports could be considerable. On the one hand, India's best talents and capabilities are employed for exporting software services, while software for domestic use is largely imported from abroad. Inadequate policy initiatives to induce firms in the manufacturing sector to harness ICT to enhance their productivity and competitiveness along with software firms to provide greater focus on domestic market as a spring board for capability building seems to have stunted the diffusion of IT into the domestic economy. For instance, the availability of software in local languages could have facilitated a widespread diffusion of IT in the country. The lost opportunity for productivity improvement through the diffusion of IT in India could be substantial. On the other hand, Indian software companies' contribution to productivity improvements in the US industry could be significant. To some extent the prevailing fiscal incentive regime that is availability of tax incentives for export profits, diverts attention towards exports by making them more rewarding activity compared to serving the domestic market. Hence there is a need for rethinking the present incentive structure to software industry (Kumar and Joseph 2005).

At present, most of the export-oriented software companies operate as "export enclaves" with little linkages with the domestic economy, if at all (D'Costa 2003, 2004). MNE subsidiaries in software development, in particular, derive almost all of their income from exports to their parents. Hence, hardly any vertical linkages are developed with the domestic software market or the rest of the economy. The enclave nature of operation generates very few knowledge spillovers for the domestic economy. The bulk of the work done is also of highly customized nature having little applications elsewhere.

Studies also analyzed the product market (spending induced real appreciation) and factor market (resource movement effect) implications of IT export boom. It was shown that the IT export boom might have had a dampening effect on the cost and competitiveness of other sectors through the resource movement effect. Sectors that seem to have had adverse effects are the ones that compete with the IT sector for skilled manpower, such as IT hardware and electronics, IT training, R&D and so forth. Thus while IT has the potential of contributing towards productivity and growth, excessive export orientation with less focus on the domestic market seems to have had adverse impact on other sectors competing for skilled manpower, at least in the short run

(Joseph and Harilal 2001, Joseph 2002, Joseph 2005d). These studies have called for greater focus on promoting IT in different sectors of the economy.

With a view to gauge the implications of software export boom for other sectors competing for skilled manpower, Joseph (2005d) estimated trends in the share of electrical, electronics and industrial machinery industry in the total investment, employment and output of the manufacturing sector in Karnataka.[7] It is found that during 1980s the share of these industries in the total industrial investment, employment and value of output increased. But in the 1990s, as the software exports have been booming, their share steadily declined and by 1999 the observed share was comparable to their share in the early years of 1980s (Figure 2.3). The analysis, though preliminary, tends to suggest that the boom in software exports has come at a cost in terms of weakening the industrial base of certain skill-intensive and key industries like industrial machinery, electrical and electronics industries.

Lagging hardware sector

India is one of the pioneering countries in the developing world to develop an electronics production base and it has a highly diversified

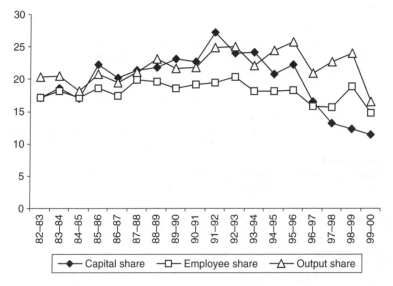

Figure 2.3 Performance of sectors competing for skilled manpower in Karnataka
Source: Joseph (2005d).

product structure with more than 3000 firms engaged in the production of a wide range of electronic consumer goods, electronic capital goods and intermediates (Joseph 1997, 2004b). However, in recent years, while the software sector has been experiencing unprecedented growth, both in exports and in production, the electronics industry in general and computer hardware production in particular, has been recording negligible and even negative growth rate (Table 2.3) with the growth rate of electronics output not even keeping pace with the overall growth of the economy.

Such an unsatisfactory performance of electronics industry in recent years needs to be seen against the fact that during the 1970s and 1980s recorded growth rate of electronics industry has been significantly higher than that of the industrial sector in general and the manufacturing industries in particular. These higher growth rates, driven mostly by domestic demand, need to be viewed along with the limited exposure of the industry to international competition. Hence the recent decline in growth rate tends to suggest that the

Table 2.3   Production and export of electronics products from India (US$ million)

Products Category	1996–97	1997–98	1998–99	1999–00	2000–01	2001–02	ACGR
Consumer electronics							
Production	1830.99	2026.67	2216.87	2604.65	2510.87	2662.47	7.77
Export	185.9	98.67	102.41	104.65	134.78	146.75	−4.62
Communication and broadcasting							
Production	845.07	866.67	1060.24	1209.3	978.26	943.4	2.22
Export	91.27	80	60.24	41.46	126.09	31.45	−19.19
Instruments and strategic electronics							
Production	1239.44	1080	1108.43	1209.3	1250	1331.24	1.44
Export	49.58	64	42.17	39.53	121.74	199.16	32.06
Computer hardware							
Production	771.83	746.67	554.22	581.4	739.13	733.75	−1.01
Export	375.49	293.33	72.29	139.53	260.87	377.36	0.09
Electronics components							
Production	1042.25	1173.33	1144.58	1209.3	1195.65	1194.97	2.77
Export	179.44	213.33	216.87	279.07	397.39	461.22	20.78
Total electronics							
Production	5729.58	5893.34	6084.34	6813.95	6673.91	6865.83	3.68
Export	881.68	749.33	493.98	604.24	1040.87	1215.94	6.64

Source: Joseph and Parayil (2005).

industrial structure that evolved over the years has not been competitive enough to withstand the new competitive pressures under globalization. A study on the firm-level production and export performance of electronics industry using a multinomial logit model has shown that the firms that managed to record higher growth in production and exports are the ones with foreign equity participation, higher investment in R&D, higher dependence on imported components, spares and capital goods and larger size. Thus the study underscores the need for creating an innovative environment by promoting R&D and other related activities along with a liberal trade and investment regime (Joseph 2005c).

ICT use: Telecommunications

In 1948 the teledensity in India was estimated at 0.02 and even after 50 years the teledensity in the country remained at a level as low as 1.94. Since 1998, however, there has been remarkable increase in teledensity and it reached 10.38 by September 2005 (Table 2.4). Table 2.4 also indicates that the observed increase has been mostly on account of the increased access to mobile telephones, and by October 2004 the number of mobile telephones exceeded the fixed telephones in the country. It has also been estimated that during 1994 almost two million new telephones were added every month and in September 2005 three million telephones were added (TRAI 2005).

Table 2.4   Growth of fixed and mobile telephones in India

Year	Fixed	Mobile	Teledensity
1995	9.80		
1996	11.98		
1997	14.54	0.34	
1998	17.80	0.88	1.94
1999	21.59	1.20	2.33
2000	26.51	1.88	2.86
2001	32.44	3.58	3.58
2002	41.48	13.00	4.28
2003	42.58	33.58	5.11
2004	45.00	50.00	7.02
2005	47.83	65.05	10.38

Note: Data for the year 1995 refers to the period up to September 2005.
Source: TRAI (2005) and Cellular Operators Association of India (2005).

Notwithstanding the dramatic increase in teledensity, thanks to the liberalized policy initiatives by the states, the intra-national digital divide as manifested in the gap between rural and urban teledensity also increased. While the teledensity in the urban areas increased from 5.78 in 1988 to over 26 in 1995 that in the rural areas increased only from 0.43 to 1.74 (Figure 2.4). This needs to be viewed against the fact that the access to mobile telephones, which played a key role in raising the teledensity in the country, has been confined mostly to the urban areas. Thus viewed, the remarkable growth in teledensity in the recent years has been mainly on account of the meeting of pent-up demand mostly from the urban areas and the rural areas are yet to get benefited from telecommunications. At this juncture it goes without saying that similar to telephones the access to Internet also remains confined mostly to the urban areas. On the whole, given the observed pattern of telecommunication growth in the country, greater attention is called for in the coming years to address the growing intra-national digital divide.

ICT diffusion in manufacturing

It is by now recognized that the use of ICT may be instrumental in enhancing efficiency, productivity and international competitiveness of the manufacturing sector but the researchers (Nath and Hazra 2002) were much concerned with the limited linkage between the ICT sector and the industry. However, in case of India, our understanding on the extent of ICT use by the manufacturing sector

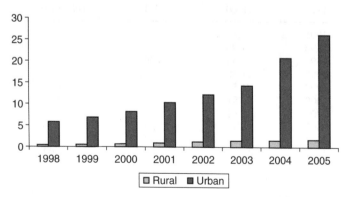

Figure 2.4   Teledensity in rural and urban areas of India

at best remains rudimentary. This is mainly on account of the fact that, as of now, there is hardly any reliable published data on ICT use by firms in the manufacturing sector. Since 1998, the Central Statistical Organization (CSO), in its Annual survey of Industries, has been collecting information on the investment in computers and software by firms in the manufacturing sector. To reflect on the issue at hand, we obtained this unpublished data from the CSO and undertook a preliminary analysis. It is found that the overall investment by the manufacturing sectors has been less than one per cent and most of the industries found spending less than 0.6 per cent (Figure 2.5). The observed level of spending needs to be viewed against the target of at least 3 per cent of the government spending set by the National Task force. At this juncture we must also hasten to add that the analysis has been preliminary and more research is called for to draw definite conclusions.

ICT and wider development issues

Realizing the potential of ICT for addressing various development issues, besides the central government, state governments have also been undertaking various ICT projects with a view to spread as widely as possible the benefits of ICT, including to the less privileged segments of society. In addition to the central and state governments, different civil society organizations have also undertaken a number of initiatives towards harnessing ICT for rural development and poverty alleviation. In addition there are a number of e-governance projects initiated at the instance of state governments and the central government.[8] The private corporate sector is also increasingly undertaking ICT projects aimed at rural development. The initiatives by Tata consultancy on literacy, HP project in the Kuppam village of Andhra Pradesh and above all e-choupal project dealing with agriculture product marketing in different states at the instance of Imperial Tobacco Company (ITC) deserves mention here.[9]

Thus, today there are different stakeholders – central and state governments, civil society organizations and the private corporate sector – involved in harnessing for addressing various developmental issues in India. While the large number of initiatives by different stakeholders have created an impression that India is well on the path to harnessing ICT for development much ahead of other developing countries, in reality they have been sporadic, involving the process of learning by doing, very often than not, lacking strategic sense and a national perspective and resulting in duplication of efforts. Also many of the

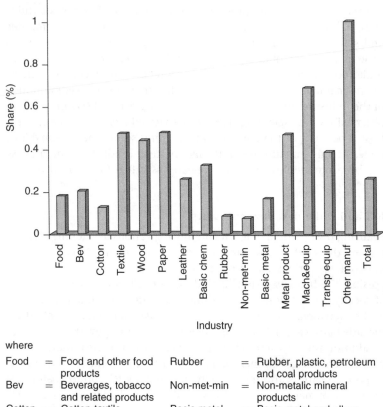

where

Food = Food and other food products

Bev = Beverages, tobacco and related products

Cotton = Cotton textile products

Textile = Textile products (including wearing apparel)

Wood = Wood and wood products

Paper = Paper and paper products

Leather = Leather and leather products

Basic chem = Basic chemicals and chemical products

Rubber = Rubber, plastic, petroleum and coal products

Non-met-min = Non-metalic mineral products

Basic metal = Basic metal and alloys industries

Metal product = Metal products, except machinery and equipment

Mach & equip = Machinery and equipment other than transport equip

Transp equip = Transport equipment and parts

Other manuf = Other manufacturing industries

Figure 2.5 Share of IT investment in total fixed investment across different industries in India (2001–02)

experiments remain at pilot stage, accomplishing much less than what could have been possible had there been a national system of innovation attuned to effectively harnessing the new technology for development.[10] On the whole, it appears that while India is known as a major producer of ICT, her progress in ICT use so far has not been as remarkable and seems to have not effectively articulated the social marginal product of a dollar worth of ICT used at home as compared to a dollar worth of ICT exported (Joseph 2002).

From the above discussion, while India's performance with respect to the production and export of IT software and services has been highly remarkable, it appears that there is room for improvement with respect to the production of ICT goods and also ICT use in different sectors of the economy. Nonetheless, as per the Network Readiness Index (NRI) – an indicator of the overall performance of a country with respect to ICT – India performed much better as compared to countries with comparable levels of per capita income. In terms of the NRI, India occupies 37th rank among 82 countries for which the Index has been constructed (Dutta *et al.* 2003). It is to be noted that other countries with comparable levels of Index were found having per capita income three to four times higher than India. At the same time, the rank of countries with per capita income comparable to India was in the range of 75–82 (Table 2.5).

Table 2.5 Network Readiness Index and rank of selected countries (2002–2003)

Name	NRI score	NRI rank	PCI
Chile	4.14	35	9,417
South Africa	3.94	36	9,401
India	*3.89*	*37*	*2,358*
Latvia	3.87	38	7,045
Poland	3.85	39	9,051
Slovak Republic	3.85	40	11,243
Ecuador	2.6	75	3,203
Paraguay	2.54	76	4,426
Bangladesh	2.53	77	1,602
Bolivia	2.47	78	2,424
Nicaragua	2.44	79	2,366
Zimbabwe	2.42	80	2,635
Honduras	2.37	81	2,453
Haiti	2.07	82	1,467

Note: NRI – Network Readiness Index. PCI – per capita income, at purchasing power parity.
Source: Dutta *et al.* (2003).

III. Behind the remarkable performance: Innovation system and the trade regime

It is clear from the discussion so far made that the performance of export-oriented IT software and service sector in India has been remarkable notwithstanding certain opportunity cost associated with it. Naturally, the remarkable performance of India has induced many developing countries to search for lessons from Indian experience. This in turn calls for a better understanding of the factors that facilitated India's performance. Some observers have tended to argue that India's success in the software and service industry has been an outcome of the free play of market and of benign state neglect (Arora *et al.* 2001). Such observations based on casual observation and common sense conceal more than what they reveal and are also of little relevance for other developing countries searching for lessons from India's experience. This is because, there are number of instances from the developing countries that resorted to liberal trade and investment regime with greater role for the market but with limited success in developing a skill- and knowledge-intensive sector like IT software and services. Therefore in this section, drawing heavily (from Kumar and Joseph 2004), we shall search for the factors that configured India's success using the broad analytical framework of NSI. We shall also explore the role that trade and investment policies played in making India an IT powerhouse of the South.

Institutional infrastructure and policy initiatives

As early as in the mid-1960s the Government of India recognized the critical importance of electronics industry and computing for national development in view of their "pervasive" applications and has consciously strived to build local institutional infrastructure for development of local capabilities. As will be seen below, these early initiatives have provided a base for rapid development of IT software industry in the 1980s and 1990s.

In contrast to the general perceptions, the importance of promoting software development, particularly for export, had been recognized by the erstwhile Department of Electronics, and suitable policies and programs were put in place as far back as 1972 (Parthasarathi and Joseph 2004). In a period when very high tariff and non-tariff barriers were the rule, import of computer systems on a custom duty-free basis and without reference to indigenous angle clearance was permitted for software export. Moreover, in a period when there were a series of restrictions on FDI, 100 per cent foreign-owned companies were permitted to set up software export operations, provided they locate in the Santacruz Electronics Export Processing Zone (India, Department of Electronics 1972).

A series of government committees and policy measures have contributed to the evolution of the NSI in the IT sector. The early initiatives include Bhabha Committee of 1963, Electronics Committee Chaired by Dr V.A. Sarabhai in 1966, and the National Conference on Electronics of March 1970.[11] Significantly, the National Conference had recognized the potential of India to emerge as a force in software exports. As a follow-up of their recommendations, a separate Department of Electronics (DoE) was set up to coordinate and implement policies for development of electronics industries including computer software in 1970. In 1971 the government constituted the Electronics Commission as a policy formulation body with a heavy emphasis on R&D and technology development. In 1973 the Technology Development Council was set up to assist the Electronics Commission on the recommendation of the National Seminar on R&D Policy in Electronics. DoE spearheaded a number of programmes of human resource development for software engineers, technological and communication infrastructure for software development and other facilitating measures over the past three decades in tune with the recommendations of the Electronics Commission's perspective plan.

The Computer Policy of 1984 gave a thrust to software development by underlining the need for institutional and policy support on a number of fronts. The policy, for example, called for the setting up of a separate Software Development Promotion Agency (SDPA) under the DoE. Imports of inputs needed for software development were made more liberal. However, the policy also emphasized that

> Effective software export promotion on a sustained basis can be effective in the long run only if it is planned as a part of an overall software promotion scheme covering both export and internal requirements including import substitution. Also planning for software development is integrally connected with the plan for hardware development and system engineering. (India, Department of Electronics 1985, p. 3)

The accelerated growth of computer industry following the above policy posed numerous problems for the software activities calling for a rationalization of the different aspects of policies that govern the local development and export of software. At the same time, world trade in computers was expected to be of the order of US$100 billion by 1990 wherein more than half was estimated as software. Against this background, there was the felt need for more concrete policies towards the promotion of software development and export. Accordingly, in 1986 an explicit software policy was announced and software was identified as one of the key sectors in India's agenda for export promotion. The policy underlined the importance of an integrated development of software for domestic and export markets (India, Department of Electronics 1986). The policy had the following specific objectives:

• to promote software exports to a take a quantum jump and capture sizeable share in international software market;
• to promote the integrated development of software in the country for domestic as well as export markets;
• to simplify the existing procedures to enable the software industry to grow at a faster pace;
• to establish a strong base of software industry in the country;
• to promote the use of computer as a tool for decision making and to increase work efficiency and to promote appropriate applications which are of development-catalyzing nature with due regard to long-term benefit of computerization to the country as a whole.

To facilitate the realization of the stated objectives, the policy emphasized the need for simplifying the existing procedures pertaining to all the aspects of software development and production for both domestic and export markets and provided various fiscal incentives to software firms, such as like tax holidays, tax exemption on the income from software exports, and duty-free import of hardware and software for 100 per cent export purposes.

With the initiation of economic reforms in the early 1990s the finance ministry made an assessment that apart from the general orientation of all industries towards export markets, India's comparative advantage was in the software and not in the hardware. Therefore, a major thrust was consciously given to software exports. Accordingly, new policy measures have been initiated which *inter alia* included removal of entry barriers for the foreign companies and removal of restrictions on foreign technology transfers, participation of the private sector in the policy making, provisions to finance software development through equity and venture capital, measures to make available faster and cheaper data communication facilities, and reduction and rationalization of taxes, duties and tariffs and so on[12] (Narayana Murthy 2000).

Along with the policy measures initiated by the national government, various state governments have also enacted IT policy with a view to promote ICT growth in the respective states. As of 2004, 18 of the state governments have enacted such policies. These policies generally focused on the key issues of infrastructure, electronic governance, IT education and provided a facilitating environment for increased investment in the IT sector of respective states.[13]

By recognizing the potential of IT-related industries and software for India's development, the prime minister appointed a National Taskforce on Information Technology and Software Development (NTITSD) in May 1998 under the chairmanship of the deputy chairman, Planning Commission. NTITSD submitted its report outlining a National IT Plan comprising 108 recommendations for the software and 87 recommendations for the hardware (India, NTITSD 1998). These recommendations have since been notified by the government in the Gazette of India dated 25 July 1998. NTITSD has set before the country an ambitious target of $50 billion software export by 2008. The department of Electronics was upgraded into a full-fledged Ministry of Information Technology (MIT) in October 1999 to coordinate the promotional role of the government in the industry.

Supply of trained manpower for software development

The government permitted private investment in IT training since the early 1980s. These privately run centers offer diplomas of various duration ranging from short-term specialized courses to longer-term basic courses. However, the quality of the training imparted by these institutions has been uneven. DoE stepped in with a scheme called Department of Electronics Accreditation Society (DOEACC), jointly with the All India Council of Technical Education (AICTE) in 1990 to provide accreditation to specified level of courses, namely, O-foundation course, A-Advanced Diploma, B-MCA Level and C-M.Tech Level. By January 2004, DOEACC Society had accredited a total of 850 institutes. The Society conducts examinations for all the four levels twice a year and grants certificates/diplomas (India, Ministry of Information Technology 2004).

The demand for software personnel, especially engineering graduates, has grown rapidly since the mid-1990s due to the expansion of the software development activity in India as well as the growing brain drain. In view of this, easing the supply of IT professionals has been one of the challenges faced by the country. The NTITSD has made a number of recommendations dealing with augmenting the quality and quantity of trained manpower for software industry. In tune with these recommendations, the capacity of the higher education system in engineering in the country has been expanded besides setting up of new institutions (Kumar 2000). English is widely used as a medium of instruction, which in turn provides an ample supply of manpower for ITES services at a much lower cost as compared to other countries. While the present supply of IT manpower is considered adequate, concern is with the quality and supply of middle-level manpower.

IT infrastructure: Software technology parks

A notable institutional intervention has been the establishment of STP[14] to provide the necessary infrastructure for software export. The first ones to come into being were those at Bangalore, Pune and Bhubaneshwar in August, October and December 1990 respectively. In 1991, four more STPs were set up by the DoE at Noida, Gandhinagar, Trivandrum and Hyderabad. By 2003 there were 39 STPs and the units registered with these STPs accounted for about 80 per cent of the exports of software from the country. The infrastructure facilities available in these STPs included, among other things, modern computers and communication network, which are beyond the reach of individual firms. The STPs also envisage a transparent policy environment and a package of concessions (India, Ministry of Information Technology 2004).

In June 2000, a new STP was set up in Silicon Valley, composed of a Business Support Centre and an India Infotech Centre, with a view to facilitate software export by small and medium firms to the United States of America.[15] The center also fosters business relationships by providing access to financial institutions in the USA, Venture capital funds and specialized trade bodies to promote partnerships and strategic alliances between American and Indian ICT software and service companies (Parthasarathi and Joseph 2002).

IT infrastructure: Reforms in the telecom sector[16]

Realizing the importance of telecommunication in overall development of the economy in general and that of IT and Software service sector in particular, the government initiated a number of policy reforms that helped create a highly competitive environment leading to drastic reduction in telecom cost and also to increased access and better quality services. Perhaps the first step was the announcement of the National Telecom Policy in 1994, opening up the telecom sector to competition in Basic Services as well as Value Added Services like Cellular Mobile Services, Radio Paging, VSAT Services and so on. It also set target for provision of telephone on demand and opening up of long-distance telephony. This was followed by the Telecom Regulatory Act of 1997, which led to the establishment of an independent Telecom Regulatory Authority of India (TRAI). Later, the New Telecom Policy of March 1999 focused on removing some of the bottlenecks and moving the liberalization process forward (ITU 2002a). These policy reforms have resulted in the very rapid expansion of the telecom network in the country. This has been possible because of opening up of all the telecom services for the private sector without any restriction on number of operators except for the cellular mobile phone segment due to frequency constraints. Private sector investment has been helping in bridging the resource gap to a considerable extent as was envisaged.

Government also permitted foreign investment in laying submarine cables, which led to the joint ventures like the one between Singapore Telecommunications Ltd. (SingTel) and Bharti Enterprises of India, a US$650 million joint venture, to build the world's largest cable network in terms of capacity. Today, submarine cables account for about 30 per cent of international long-distance dialing, and with additional investment in the pipe line the share is expected to increase in the near future.

Recent initiatives include permission for offering Internet telephony, opening up of National Long Distance Service (NLDS) to private operators by abolishing the monopoly of VSNL in International Long Distance Service. Along with these policy measures, a liberal policy for Internet Service Provider was also announced. As of December 2002, licenses were provided to 573 ISPs of which 24 have been given clearance for commissioning of 55 international gateways for Internet using satellite media (India, Department of Telecommunications 2004).

As a result of the liberalization in the policy towards FDI in general and foreign investment in the field of IT and telecommunications, between August 1991 and December 2002, 858 proposals of FDI amounting to Rs. 5627.9 million were approved and the actual flow of FDI during the above period was Rs. 956.2 million. In terms of approval of FDI, the telecom sector is the second largest after the power and oil refinery sector. In 2002 (January–December), the actual FDI inflow was of the order of Rs. 108.1 million.

R&D capability building

Department of Electronics has put heavy emphasis on R&D activity relating to, among other fields, development of computer software by supporting R&D activity in the area at different institutions such as Tata Institute of Fundamental Research (TIFR), Indian Institute of Technology (IIT), Indian Institute of Science (IISc), select universities (such as Jadavpur University) and Council of Scientific and Industrial Research (CSIR) Laboratories since the early 1970s. The Technology Development Council has been supporting R&D projects since its inception in 1973 (India, Ministry of Information Technology 2004). These programmes of technology development have led to building up of capabilities and have provided experienced manpower for the rapid development of the industry in the coming years. For instance, the capabilities built in the process of early work on data communication at TIFR started in the late 1970s and anchored at the DoE supported National Centre for Software Technology (NCST), set up in Mumbai in 1984, proved instrumental for the development of country-wide networks and for Internet in the country in the 1990s. The government S&T agencies have set up a parallel Super-computer Education and Research Centre (SERC) and Department of Computer Science and Automation at IISc, which provided high-end expertise and manpower to the industry in software. Besides NCST, DoE has also set up another institution for technology development in the 1980s, namely Centre for Development of Advanced Computing (C-DAC). C-DAC has developed India's first super computer, Param, and has developed software for Indian languages' script. Electronics Research and Development Centre (ER&DC) set up by the DoE also plays an important role in R&D in IT and electronics. The government has also stimulated and supported R&D activity of industry through tax incentives and direct funding on a limited scale by DoE (Kumar 2001b).

Institutional infrastructure and patterns of spatial agglomeration

Information Technology and software development in different parts of the world is characterized by a strong tendency of clustering because of agglomeration economies. In India the software industry developed initially in Mumbai (formerly Bombay) followed by concentration in Bangalore, New Delhi, Hyderabad and Chennai. As we have already seen, the top five cities together account for the bulk of total exports from India. The pattern of concentration of software development industry in and around select cities does corroborate the key importance of institutional infrastructure for the activity. Table 2.6 shows

Table 2.6 Illustrative S&T infrastructure in four IT clusters in India

Type of NIS infrastructure	Bombay	Bangalore	Delhi and around	Hyderabad
Institutions of higher technical education and excellence	IIT-B; Bombay University; SNDT Women's University; Bajaj Institute of Management and several other engineering and management institutes	IISc; University Visvesraya College of Engineering; SKSJ Technology Institute; and 28 private engineering colleges; Indian Institute of Management-B	IIT-D; Delhi College of Engineering; Delhi University Department of Computer Sciences; Roorkee University of Engineering (within 200 km); J.N. University; Jamia Milia Islamia Engineering College; FMS; IIFT; plus several private institutions	J.N. Technological University; Hyderabad University; Osmania University; Kakatiya University;
Public-funded research laboratories and institutions	TIFR; NCST; BARC; UDCT; SAMEER	ISRO; NAL, CMTI; Electronics and Radar Development Establishment Aeronautical Development Establishment; Gas Turbine Research Establishment; Centre for Aeronautical Systems Studies and Analysis; ER&DCI	NIC; NPL; Institute for Systems Studies and Analysis; SPL; C-DOT	National Remote Sensing Agency; RRL; NGRI; IICT; Defence Electronic Research Laboratory; DRDL

Table 2.6 (Continued)

Type of NIS infrastructure	Bombay	Bangalore	Delhi and around	Hyderabad
Local software champions	TCS; PCS; Tata Infotech; Mastek; L&T ITL; APTECH; COSL; Datamatics; Silverline	Infosys Technologies Ltd; WIPRO Information Technologies	HCL Technologies; NIIT Ltd; CMC Ltd;	Satyam Computer Services Ltd
High-speed data communication facilities	Earth Station of STPI	Earth Station of STPI	Earth Station of STPI	Earth Station of STPI
High technology enterprises, mostly public sector	L&T; Godrej; Tata group and a large number of engineering and electronics enterprises	ITI; BEL; HAL	Central Electronics Ltd; NRDC; EIL; RITES; ETTDC; ET&T; RITES; TCIL	ECIL; BHEL

Source: Kumar (2001b).

that the cities of high concentration of software development – namely Mumbai, Bangalore, Delhi and Hyderabad – have shared a disproportionate share of national innovative infrastructure, skill base and other resources for technology development. Because of significant agglomeration economies present in skill- and knowledge-intensive activities such as software development, this disproportionate share of national innovative infrastructure has crowded in the software development activity in these cities.

Role of industry associations

It may be myopic to attribute the observed dynamism of the industry entirely to the initiatives made by the state (Joseph 2002). There were also other factors which include, but not limited to, the availability of highly skilled English-speaking labor force at a wage rate much lower than that of the developed countries, time difference between India and the US, which still continues to be the major export market. While the state initiatives laid the foundation for faster growth, the industry associations,[17] particularly the NASSCOM played an important role. In addition to lobbying at the central and state government levels, the NASSCOM also played a key role in projecting India's image in the world IT market. For example, in 1993 NASSCOM appointed a full-time lobbying firm in Washington. NASSCOM also facilitated the participation of Indian firms in a large number of international IT exhibitions and thus projecting India's capabilities in the sphere of IT. The role that NASSCOM played in getting the visa rules relaxed by the developed countries, especially the USA, is well known. Also, in 1994 NASSCOM initiated the antipiracy initiatives in India, when IPR was becoming a major issue in the Indo-US relations. It took up the campaign against software piracy and conducted a number of well-publicized raids.[18]

Trade and investment policies

Studies (Joseph 1997, 2005c) have identified three phases in the evolution of trade, investment and other policies governing the growth of Electronics and ICT industry in India. The main anchor of the first policy phase, covering the 1970s, was around the development of the industry under protection with minimal recourse to foreign capital and foreign technology on the one hand and large companies and Business Houses on the other. The 1980s marked the second phase when the government took a number of initiatives in the direction of a more liberal and open policy regime. Broadly speaking, the shift was from the earlier controls towards a more liberal policy with the emphasis laid on minimum viable capacity, scale economies, easier access to foreign technology and capital and free entry to private sector capital, including companies covered under Monopolies and Restrictive Trade Practice Act (MRTPA) and Foreign Exchange Regulation Act (FERA) with a view to making the industry technologically modern, cost-effective and price-competitive. In a sense, the focus has been more on internal liberalization with the removal of restrictions on industrial licensing and capacity creation with a view to achieve economies of scale.

The new industrial policy of July 1991 set the beginning of the third phase marked by further liberalization in industrial licensing and greater outward orientation. Thus, 1990s witnessed the removal of industrial licensing for most of the products except a few products of strategic significance and further liberalization with respect to FDI and technology import along with a series of fiscal and trade policy reforms to facilitate production and outward orientation. The policy reforms were based on the basic premise that the trade and investment provides an enabling environment for the overall development of an economy in general and developing an industrial base, especially that of high-tech industries, in particular and that there is an *inter-se* link between the two.

The new industrial policy divided the foreign investment proposal into two categories: those cleared by the RBI (automatic approval) and those cleared by the Foreign Investment Promotion Board (FIPB). The FDI policy provides for automatic approval by the RBI if the foreign equity is less than 50 per cent, lump sum payment for the price of technology not exceeding $2 million, royalty payments not exceeding 5 per cent of domestic sales and 8 per cent for exports (net of taxes).

The Foreign Investment Promotion Board cleared the proposals that did not confirm to the guidelines of automatic approvals. Government also encouraged investment from NRIs including Overseas Corporate Bodies (OCBs) to complement and supplement domestic investment. Investments and returns were made freely repatriable, except where the approval is subject to specific conditions such as lock-in-period on original investment, dividend cap, foreign exchange neutrality, and so on as per the notified sectoral policy. The condition of dividend balancing that was applicable to FDI in 22 specified consumer goods industry were withdrawn for dividends declared after 14 July 2000.

In a major drive to simplify procedures for FDI under the "automatic route", RBI has given permission to Indian companies to accept investment under this route without obtaining prior approval from RBI. Investors are required to notify the regional office concerned of the RBI of receipt of inward remittances within 30 days of such receipt and file required documentation within 30 days of issue of shares to foreign investors. This facility is available to NRI/OCB investment also.

The salient features of the investment policy applicable to Electronics/IT and Telecom Sector are as follows:

1. Automatic route for foreign equity up to 100 per cent in software development and Electronics/IT/Telecom hardware manufacturing, except aerospace and defence and small scale reserved items.
2. 100 per cent foreign investment permitted in units set up exclusively for export. Such units can be set up under any of the following schemes, namely, Electronics Hardware Technology Park (EHTP), Software Technology Park (STP), Export Oriented Units (EOUs), Special Economic Zones (SEZs).
3. In basic, cellular, value added services and global mobile personal communications by satellite, FDI is limited to 49 per cent (which has

been raised to 74 per cent in 2004–2005 budget) subject to licensing and security requirements and adherence by the companies to the license conditions for foreign equity cap and lock-in period for transfer and addition of equity and other license provisions.

4. In Internet Service providers (ISPs) with gateways, radio-paging and end-to-end bandwidth, FDI permitted up to 74 per cent, with FDI beyond 49 per cent requiring Government Approval.

5. FDI up to 100 per cent is allowed for the following activities in the telecom sector:

   a. ISPs not providing gateways (both for satellite and submarine cables);
   b. Infrastructure providers providing dark fibre (IP Category 1);
   c. Electronic Mail; and
   d. Voice Mail

   The above would be subject to the following conditions:

   a. FDI up to 100 per cent is allowed subject to the condition that such companies would divest 26 per cent of their equity in favor of the Indian public in five years, if these companies were listed in other parts of the world.
   b. The above services would be subject to licensing and security requirements, wherever required.
   c. Proposals for FDI beyond 49 per cent shall be considered by FIPB on case-to-case basis.

6. FDI upto 100 per cent is also permitted for e-commerce activities, under the Government route, subject to the conditions that such companies would divest 26 per cent of their equity in favor of the Indian public in five years, if these companies were listed in other parts of the world. Such companies would engage only in business to business (B2B) e-commerce and not in retail trading (India, Ministry of Information Technology 2004).

Foreign technology agreements

The RBI through its regional offices accords automatic approval for foreign technology collaboration agreements in all areas of electronics and information technology, except electronic aerospace and defense equipment and small-scale reserved items, subject to: (i) the lump sum payments not exceeding US$2 million; (ii) royalty payable being limited to 5 per cent for domestic sales and 8 per cent for exports, subject to a total payment of 8 per cent on sales over a 10-year period; and (iii) the period for payment of royalty not exceeding seven years from the date of commencement of commercial production, or 10 years from the date of agreement, whichever is earlier.[19]

From January 1991 to December 2002, Government approved 23,462 foreign collaborations (Technical and Financial) proposals with a corresponding FDI of US$76.61 billion and the total FDI inflows add up to US$32.41 billion. Out of these, the number of approvals for electrical equipment sector, including electronics, has been of the order of 5033 (21.45 per cent of the total approvals) with an equity participation of US$7053.0 million, accounting for about 9.82 per cent of the total investment. The electrical and electronics equipment sector ranked 3rd in the list of sectors in terms of cumulative FDI approved from August 1991 to December 2002. Of these, nearly 65 per cent of the investment has been in the field of software (Table 2.7).

Trade and fiscal policies

To promote export, special schemes have been made available under export-oriented unit scheme, electronics hardware technology park scheme, software technology park scheme and the provision for special economic zones. The units operating under these schemes are entitled to various tax incentives, which *inter alia* include access to duty free imports along with 100 per cent foreign equity (see for details, India, Ministry of Information Technology 2004).

Table 2.7 Number of foreign collaborations and FDI (approvals) in the electrical and electronics and software sectors in India during 1991–2002

Industry	No. of approvals			Amount of FDI approved		Share in sectoral total (%)
	Total	Technical	Financial	Rs. million	US$ million	
Electrical equipments	1,712	909	803	60,799.1	1,659.1	21.73
Computer software	2,707	89	2,618	181,143.5	4,382.1	64.75
Electronics	523	160	363	32,810.4	900.9	11.73
Computer hardware	22	2	20	3,785.2	79.1	1.35
Others	69	20	49	1,238.4	31.8	0.44
Total of above	5,033	1,180	3,853	279,776.6	7,053.0	100

Source: India Investment Center (2002).

To facilitate the free inflow of inputs into and output out of the country, the trade policy regimes were subjected to substantial liberalization. The first round of tariff reduction has been effected with the New Industrial policy of 1991 and further reductions were effected in the Exim policies that followed. India, being a signatory to the Information Technology Agreement of WTO, is committed to reduce the tariff on product to zero by April 2005. Physical controls on import of most of the electronic equipment and components have been by now done away with. In the budget for 2003–2004, the peak rate of customs duty has been reduced from 30 to 25 per cent. Further reduction in the duty structure has been brought about in the budget for 2004–2005. As already noted, there are also income tax exemptions on profits earned through exports for export-oriented units operating from EHTPs, SEZs and STPs. Thus viewed, the liberalized trade and investment policies along with series of fiscal concessions also played their role as these initiatives were preceded by the building up of a broad-based innovation system to facilitate the growth and diffusion of a skill-intensive technology like ICT.

IV. Concluding observations

The remarkable performance of India in the IT software and the service sector seems to have no parallels in the developing world. India can be proud of her track record in the export of IT software and services along with unprecedented growth in the access to telecommunication services and the large number of ICT projects undertaken across the country at the instance of different stakeholders to address various development issues. These achievements got manifested in her 37th rank in NRI that was estimated for 82 countries. Surprisingly, other countries with comparable levels of Index were found having per capita income three to four times higher than India. These achievements notwithstanding, the present study noted that her performance with respect to ICT use leaves room for much improvement. Also there has been certain opportunity costs associated with the export performance and there appears to be the need for inducing the IT firms to give greater focus on the domestic market. Similar to the experience of many other developing countries, there is the need for more focused efforts to address the intra-national digital divide.

Our analysis of the contributory factors towards the development of ICT sector in India has shown that the NSI evolved over the years played a significant role in making India an ICT powerhouse from the South. The institutional interventions included development of a system of higher education in engineering and technical disciplines, creation of an institutional infrastructure for S&T policy making and implementation, setting up STPs and building centers of excellence and numerous other institutions for technology development. The patterns of clustering of the software development activity in select centers provide a further evidence to the contention that public-funded technological infrastructure has crowded in the investments from private sector and foreign firms in skill-intensive activities such as software development. The government has also undertaken a series of policy initiatives as early as the 1970s, such as the liberal policy towards import of needed capital goods and a kind of open-door policy towards FDI meant for IT and software exports. Nonetheless, the returns to these initiatives remained rather limited until India resorted to a much liberal trade and investment regime, indeed at her own pace and her own terms, in the 1990s. Thus viewed, there is lot more merit in the argument that import substitution policies in India, directly and indirectly, were responsible for the emergence of a vibrant ICT sector (Patibandla *et al.* 2000). In general, India's IT success is a typical case of proactive state intervention wherein the government laid a strong foundation and created the facilitating environment, and the industry took off with greater participation by the private sector and increased world demand in the 1990s.

Having examined India's experience with respect to ICT and highlighted the broad contours of its performance and the underlying factors, let us now proceed with our exploration of the experience of another country that has adopted an entirely different route towards development. This enquiry forms the focus of the next chapter wherein we take up the case of Thailand.

3
Thailand: From Investment-led Growth to Innovation-led Growth

I. Introduction

The development strategy with focus on private investment in Thailand could be traced back to the first National Economic and Social Development Plan (1961–66), which underscored the role of private sector in industry and commerce. It was laid down that the government would operate only in those areas where the private sector was not equipped, such as the development of infrastructure. This approach towards private investment continued in the subsequent plans and later the liberal approach was extended to foreign capital. Thus, Thailand has a longer history of liberal trade and investment policies as compared to ASEAN New Comers like Cambodia, Laos, Myanmar and Vietnam. This development strategy is shown to have enabled the Thai economy to record remarkable growth rate (Jitsuchon and Sussangkarn 1993, Jitsuchon 1994) as compared to many other developing countries and not to speak of ASEAN New Comers. During 1960–73, for example, the Thai economy recorded an annual average growth rate of over 8 per cent. Though the recorded growth rate in the following decade (1974–85) was slightly lower (6.3 per cent), there was a marked revival during 1986–96 with an annual average growth rate of over 9 per cent without any parallels in the rest of the world. This was followed by the financial crisis resulting in a negative growth rate (−0.9 per cent) during 1997–2000. With an annual growth rate of over 5 per cent during 2002 and 2003 the economy has set the stage for revival wherein the electronic and electrical parts played an important role (UNESCAP 2004).

Along with high growth of the economy there has also been significant structural transformation. The share of agriculture in GDP declined from 12 per cent in 1990 to 9 per cent in 2002. While the share of agriculture declined considerably, that of industrial sector increased from 37 per cent in 1990 to reach a level of 43 per cent in 2002 and that of services declined marginally from 50 to 48 per cent. More importantly, the manufacturing sector accounted for about 34 per cent of GDP in 2002, an increase of 7 per cent since 1990. The high growth of Thai economy, therefore, has been driven by the industrial sector, more specifically the manufacturing sector. Thus, by the time IT revolution was taking root in the 1990s, Thailand has had an economy that was integrated with the world economy and highly vibrant in output growth. Also going by the available statistics, Thailand stood head and shoulders above the ASEAN New Comers in terms of ICT production and use. Against this background, in this chapter we shall explore the ICT production and use in Thailand and the role of innovation system and trade regime with a view to draw lessons for the ASEAN New Comers.

II. Present state of ICT production

ICT hardware (electronics) production

Though electronics production in the country dates back to 1960s, Thailand actively promoted electronics industry only during the late 1970s. While the number of firms that entered electronics production were only 6 during 1960–73 and 29 during 1974–79, the number increased to 211 during 1987–92 and further to 375 in 1993, reflecting the success of the government policy towards attracting investment into this sector. By the end of 2003, more than 600 firms were active in the field of electronics production generating more than 0.3 million employment. Out of the total employment in the electronics industry, consumer electronics accounted for about 19 per cent, computers and other electronics capital goods 28 per cent and electronic parts and integrated Circuits (ICs) 53 per cent. The major products of Thai electronics industry include computers and components, ICs and parts, and Hard Disk Drives and parts. In each of these product lines the world leaders have their presence in Thailand and most of them are expanding their production base. To list a few, foreign firms like IBM, Fujitsu[1] Cannon, Cal Comp, and Oki, Seagate[2], Lucent[3], AMD[4], and NS Electronics[5] have established production facilities in Thailand.

Given the specificities of electronics production, no country can afford to produce all the components and raw materials needed for electronics production. Hence, to promote investment, the trade regime needs to be one that facilitates the free flow of inputs into and output out of the economy. Data presented in Table 3.1 suggest that Thailand presents a typical case wherein the trade policy facilitated free imports of components and materials through the liberalized trade policy regime which in turn led to substantial investment on the one hand and export earning and employment generation on the other. Table 3.1 also shows that total exports from the country increased from $10 billion in 1997 to over $15 billion in 2000 and thereafter showed a declining trend. At the same time the imports also increased. Yet, the trade balance ranged from US$3 billion to US$4 billion. This data, however, does not take into account the imports of capital goods needed for electronics production, which has been estimated at about $0.8 billion.

Challenges of Thai ICT sector

Electronics industry played a significant role in the Thai economy during the past two decades, in terms of generating employment and export

Table 3.1 Structure of exports and imports in Thai electronics industry

	1997	1998	1999	2000	2001	2002
Exports (US$ in million)						
Computer and parts	6,520.92	7,575.09	7,672.38	8,432.30	7,725.83	7,282.63
Of which parts	2,973.32	4,714.60	5,861.36	6,434.79	5,941.60	4,461.88
Others	1,227.43	1,205.35	1,177.67	1,629.38	1,646.78	1,883.19
ICs and parts	2,307.93	2,518.82	2,255.00	4,464.32	3,486.71	3,439.28
Of which parts	304.93	481.98	492.69	556.52	277.50	364.95
Telecommunication equipment	505.68	528.37	532.91	884.71	796.26	875.74
Television receivers	1,028.49	1,079.58	905.80	1,093.87	915.86	1,122.86
Radio receivers	52.84	111.07	300.85	419.07	374.30	439.72
Total exports	10,415.86	11,812.94	11,666.94	15,294.27	13,298.96	13,160.23
Imports (US$ in million)						
Computer	661.72	386.23	337.67	619.80	1,098.55	1,823.81
Computer components	3,098.90	2,406.82	1,901.60	3,053.93	2,894.10	1,851.65
ICs	2,135.12	2,148.89	1,820.19	2,786.50	2,690.05	2,648.05
ICs components	2,720.24	2,802.96	2,279.98	3,673.01	3,046.62	2,824.16
Telcom equipment	1,821.00	1,350.00	1,420.00	1,100.00	940.00	1,010.00
Total imports	10,436.98	9,094.90	7,759.44	11,233.24	10,669.31	10,157.67
Trade balance	−21.12	2,718.04	3,907.50	4,061.03	2,629.65	3,002.56

Note: Data on Telecommunication imports are from US Department of Commerce (2004), Country Commercial Guide, Thailand, http://www.usatrade.gov/website/ccg.nsf/ShowCCG?OpenForm&Country=THAILAND.
Source: Bank of Thailand (2003).

earnings and thus aiding overall development. Based on a detailed analysis of the electronics industries in Southeast Asia, Ernst (2001) expressed the view that export-oriented production can no longer guarantee sustained growth and welfare improvement. Export-led production also faces serious external challenges from volatile global finance, currency and export markets. Three potential weaknesses identified by Ernst – a sticky specialization on exportable "commodities", a narrow domestic knowledge base leading to limited industrial upgrading and limited backward and forward linkages – in the context of Southeast Asian countries are also relevant to a great extent in the case of Thailand.

Electronics industry comprises a large number of products and Thailand has specialized in the mass production of a few products mainly for the export market. This has led to a kind of sticky specialization with limited backward and forward linkages, especially for materials and production equipment that give rise to very high level of import dependence and limited value addition. Mephokee (2003) notes that Thailand's ICT-related firms play a small subcontracting role by supplying minor components for foreign firms in the ICT industry. There are four main reasons for the firms to import these components from foreign suppliers: First, the production technology belongs to foreign parent companies. Secondly, there are no domestic components because the production technology is not available in Thailand. Thirdly, the quality of domestic components cannot meet the foreign company's requirements. Finally, it is easier to deal with foreign suppliers with whom long-term relationship has already been established. Thus the study concludes that Thai firms have small room to play in the Thai ICT industry.

A striking feature which can partly be attributed to the strategy being adopted is the mismatch between local production and consumption at both the component and the equipment levels. To illustrate, in the case of telecommunication equipment, Thailand exports almost 70 per cent of its production and at the same time imports more than 70 per cent of its domestic demand (US Department of Commerce 2004). The case with semiconductor devices also appears not different. The narrow production base with export orientation also has the effect of making the industry highly susceptible to international market fluctuations.

The problem got compounded with limited industrial upgrading and innovation. Following Hobday (1994), three stages could be identified in the evolution of electronics industry in Southeast Asia (Table 3.2). Most of the East Asian countries have started as Original Equipment Manufactures (OEMs). The OEM system enabled firms to export to international markets and to acquire foreign technology. In the 1980s, ODMs

Table 3.2 Stages of latecomer development

Period stage	Technological transition	Market transition
1960s/1970s OEM	Local firm learns assembly process for standard simple goods	Foreign MNC/buyers designs, brands and distributes. Also gains non manufacturing value added
1980s ODM	Local firm learns process engineering and detailed product design skills	As with above, MNC buys brands and distributes. MNC gains non manufacturing value added
1990s OBM	Local firm conducts manufacturing product design and R&D for new products	Local firm has own brand, organizes distribution and captures all value added

Source: Hobday (1994).

emerged out of OEM, as noted by Johnstone (1989) in the case of Taipei, China. In the 1990s, some of the leading firms in East Asia began their own-brand manufacture implying an upward mobility in the product value chain by becoming OBM, competing directly with international suppliers from Japan, the US, and Europe. Under OBM, the latecomer firms carry out all of the stages of production and innovation, including manufacturing, new product designs and R&D.

The issue is how Thailand fared in this respect. If the available evidence is any indication, the transition has been rather slow. It is further evident from Table 3.3 that presents the stages of development of electronics industry in Thailand and some of the other ASEAN countries.

Table 3.3 Technological stages in Southeast Asia's electronics industry

Decade	Singapore	Malaysia	Thailand	Indonesia	Viet Nam
1960s	Assembly				
1970s	Process engineering	Assembly	Assembly		
1980s	Product development	Process engineering	Assembly	Assembly	Assembly
1990s	Research and development	Product development	Process engineering	Process engineering	Assembly

Source: Hobday (2002).

It appears that the electronics industry in Thailand has been locked up in low value-adding assembly of electronics commodities, and the transition to ODM and OBM status has been rather slow which in turn points towards limited industrial upgrading.

Weak software sector

The limited industrial upgrading, perhaps, is most manifested in the limited success of Thailand in highly skill-intensive areas like software and services. Today Computer software demand in Thailand is being driven by the economy's continuing adjustment to economic crisis of 1997–98, including restructuring of business operations, expansion of operations of MNCs, increasing competition in manufacturing and distribution industries and growing use of the Internet and e-commerce. According to US Department of Commerce (2004), the total software market in Thailand has recorded a growth rate of 33 per cent during 2003–2004 to reach a level of approximately $448 million in 2004 and is projected to reach a level of $2.5 billion by 2009. In 2004 application solutions have a share of $240 million followed by application tools at $128 million. Systems software/utilities accounted for the smallest share of $8 million. Growing domestic demand and the presence of over 500 firms in 2004 (as claimed by the industry associations) notwithstanding, the domestic software production is still in its infant stage and therefore more than 70 per cent of the demand is being met by imports. It has been discerned from industry circles that Thailand has great potential in animation multimedia and other related software.

The observed trends need to be seen against the fact that Thailand entered 1990s with a relatively poorly educated workforce wherein 84 per cent of the workforce were having only elementary or lower level of education. This in turn has to be viewed in the context of the NSI evolved in the country over the years (Intrakumnerd *et al.* 2002). Lauridsen (2002) argues that technology policy came on the agenda quite late in Thailand. During the long period of stable growth (1960–86), industrial upgrading and technological development were not the prime concern of policy makers. Technology was, to a large extent, embodied in the imported machinery and equipment which – in the case of complex technologies – could be either acquired from the parent company (in case of joint ventures) or imported as turn key projects. During the golden decade (1987–97), Thailand experienced an economic growth that to a great extent was linked to exceptionally high growth rates in manufacturing exports. During the same period there was an even stronger increase in imports. This in turn reflected a weak engineering base and lack of supporting industries. Here it needs to be noted that FDI, which acted as vehicle of growth, was not subjected to local content requirement or under any pressure to develop supporting industries. It has also been argued that the subsidiaries and joint ventures created by FDI engaged in very limited training at levels above the most basic operating skills. They also undertook little or no technology development, consequently investing little in building the capabilities for such activities (Dahlman and Brimble 1990).

With respect to technology-generating efforts, a study by Arnold *et al.* (2000) which analyzed different aspects of innovation system in Thailand, *inter alia*, has shown that, during the 1980s and early 1990s, overall expenditure on R&D increased (at current prices), but it grew more slowly than the economy in general, and consequently there has been a steady decline in R&D as a proportion of GDP – falling from 0.21 per cent in 1987 to 0.12 per cent in 1996. The situation seems to have changed little till recently (Table 3.4). In this respect, the path followed by Thailand has been significantly at variance with several other countries in the region where R&D expenditure was a rising proportion of GDP. In Thailand, the proportion of total R&D performed in business enterprises was very low – estimated at around 10 per cent in 1996. Several constraints of R&D activities, mainly related to institutional and governmental support and the availability of human capital and support services, have been identified (Brooker group 2001). To what extent the above challenges are being addressed? This takes us to an examination of the recent policy reforms and institutional interventions in the field of ICT and software with focus on building up an innovation system.

Table 3.4 GERD and GERD/GDP of Thailand and other countries (2001)

Economic level	Development	GERD (US$ million)	GERD/GDP (%)
Advanced	Japan	121,250	3.17
countries	USA	243,548	2.69
First-tier NIEs	Korea	10,028	2.47
	Taiwan	5,903	2.05
	Singapore	1,641	1.89
Second-tier	Malaysia	296	0.39
NIEs	Thailand	269	0.26

Source: Intarakumnerd and Panthawi (2003).

III. Addressing challenges: Focus on innovation system

Thailand is one of the pioneering countries in the developing world to devise an institutional and policy environment for the development and harnessing of IT for development. Perhaps the first explicit initiative in this direction could be traced back to March 1992, when the National Information Technology Committee (NITC) was formed which was chaired by the prime minister, and comprised of ministers, permanent secretaries and high-level officers from both public and private sector with National Electronics and Computer Technology Center (NECTEC) as its secretariat. The committee was entrusted with the task of facilitating the development of ICT and promoting the use of the same in different sectors of the economy such that the new technology acts as an instrument of socio-economic transformation. By the time the MIT was formed in October 2002, the NITC had formulated three major policies pertaining to ICT in Thailand, namely IT-2000[6] – the first national IT policy of Thailand – and subsequently IT-2010, a vision document providing a long-term perspective and the IT Master Plan for the period 2002–06.

IT-2000: The first national IT policy of Thailand

IT–2000, the first national IT policy of Thailand formulated in 1995, focused on the creation of the *fundamentals* for ICT development and use on the one hand and on making use of ICT for good governance on the other. Hence, the three pillars of IT 2000 were; Creating a National Information Infrastructure (NII), Investment in human resources and improvement of public services via public electronic information.[7]

Creation of IT infrastructure

Given the highly unequal distribution of IT infrastructure, both across different regions and sections of the society, emphasis was on developing an equitable information infrastructure. The policy was unequivocal when it stated "While the statistics may look impressive, there will be about one telephone for 10 persons by 1996 and one for every 5 persons by 2001, in reality, only about a third of the population residing in the metropolis and large cities will benefit from it. Many *tambons* and almost 6000 plus villages are still be without public phones" (National Information Technology Committee 1996, p. 3). Accordingly, in the telecommunication expansion program, which envisaged an annual investment of 6000 million Baht per year during 1996–2000, emphasis was given to rural areas with a view to enhance the access to rural people. In more precise terms, the policy set the target of laying telephone lines in at least

12,000 remote *tambons* and at least one public telephone booth for every village. With a view to provide access at reasonable prices, the policy envisaged the need for a more competitive environment in the telecom sector. Hence, it underscored the need for reforming the institutional arrangements by setting up an independent Telecommunication Regulatory Authority for the regulation of different actors and also for reforms in the Telecommunication Act in tune with changing demand conditions.

Human capital

Human capital plays a dual role in IT – as producer and as user. Hence, the policy laid emphasis on increasing the supply of IT manpower at all levels. The human resource development program for IT got manifested in providing computer access to schools and universities and also in the networking of all the universities, colleges and schools. In this context the policy envisaged the setting up of a National Interactive Multimedia Institute that was expected not only to design courses but also to develop needed software.

IT for governance

The policy also envisaged making use of IT for ensuring good governance. With this strategic objective, the policy envisaged informatization of governance at all levels in the country. Also it was envisaged that at least 3 per cent of the budget on personal expenditure was to be spent on IT and of which at least 75 per cent to be spent on creating IT infrastructure including hardware, software, networking and so on. Other policy directions included providing support for the development of a strong local information industry with the active role for the private sector in all aspects of IT development as manifested in the setting up of the Software Park of Thailand (Box 3.1).

Results of an evaluation of IT-2000 undertaken by an independent agency, as quoted in Intarakumnerd and Panthawi (2003) indicated that Thailand made significant progress with respect to information infrastructure. Telephone penetration in the country increased tremendously by expanding coverage to all tambons around the country and by making available public telephones in all the villages. With respect to human resource development there has also been substantial progress. By the end of 1998, the computer–student ratio was 1:84 for primary-school level and 1:53 for secondary-school level against the target of 1:80 and 1:50 set by IT 2000. Also, by the end of 2000, almost all universities were connected to ThaiSarn, while more than 3000 schools were connected to SchoolNet.

Box 3.1   Major software initiative: The software park of Thailand

While Thailand is yet to make a significant entry into the exports of Software, in 1997, the National Science and Technology Development Agency (NSTDA), the largest Research and Technology Organisation (RTO) in Thailand, received an approval by the cabinet to set up the Software Park of Thailand (SPT) with a built up space of 13,000 square meters in order to induce the first local cluster of software industries. It received strong support from well-known transnational corporations like IBM, HP, SUN, Oracle and others. Around 50 companies, mostly Thai nationals, have fully occupied the available space. Among various facilities, NSTDA together with the Carnegie-Mellon University has consistently offered training and certifying the Capability Maturity Model (CMM) to raise the standard of software firms. Many of these firms now have customers and business links with their foreign counterparts in US, EU, Australia, New Zealand, Malaysia and so on. Local universities are also participating with some companies in the SPT to produce the local e-learning services. The establishment of SPT has facilitated technology transfer within and outside the Park and encouraged a first step toward the clustering concept of innovation that the firms will be able to learn from each other, NSTDA, participating universities, and firms located outside the park, especially transnational corporations (Joseph and Intarak-umnerd 2004).

In sum, since the initiation of IT-2000, Thailand has been successful in creating a facilitating environment for promoting the production and use of IT. However, as will be evident from later discussion, the achievements with respect to human resource development, creation of a more competitive environment in telecommunication, and the development of IT production, especially software, were less pronounced. Here it may also be noted that the policy was drafted during a period wherein the economy has been going through the high growth decade, whereas the economy experienced the financial crisis as the plan got implemented. Yet in many aspects the achievements were above the targets like in the case of the provision of computers to schools. The slip between the targets and achievement in some of the areas, perhaps, needs to be seen in the context of difficult economic conditions that prevailed. It is against this background that the IT vision for 2010 has been formulated and being implemented.

Towards the future: IT-2010

IT-2010, a national IT policy framework governing a 10-year period or a vision document, was drafted and approved by the Cabinet in March 2002. While IT-2000 focused on the IT infrastructure, IT-2010 extended the focus to include not only further strengthening of the IT infrastructure but also application domains in which IT should be utilized. More importantly, the focus of IT-2010 has been on building up an NSI to promote effective use of IT to facilitate the transition of Thailand into a knowledge-based economy and society, wherein creation, collection, dissemination and utilization of knowledge are central to economic and social development.

IT Master Plan (2002–2006)

In addition to IT-2010, NITC also drafted a five-year plan called National ICT Master Plan 2001–2006 with the top priority to ICT industry development (software industry in particular), human capital development, ICT utilization within the public sector through e-governance and strengthening the NSI.

With respect to the IT industry, the Master Plan has a number of laudable targets. This includes taking software output to 90 billion Baht by 2006 with 75 per cent exports. In addition it aims at promoting open source software with the value of at least 50 per cent of the total software market. By 2006, the Plan aims at raising the proportion of employment within ICT-based industries to a level of at least 600,000 persons (or 1 per cent of total national workforce). In addition it has the target to increase the market value contributed by e-commerce at a minimum rate of 20 per cent annually and by 2006, the economic contribution of ICT-based industries to reach at least 10 per cent of the total national economy.

In terms of developing IT manpower, the targets include (i) at least 70 per cent of the workforce should have an access to ICT and 40 per cent should have an access to the Internet; (ii) software manpower outturn of 60,000, with 30 per cent certified software developers; (iii) at least 90 per cent of all students should be ICT literate; and (iv) the number of knowledge-workers should be increased by at least 150,000 persons.

With respect to developing an NSI in ICT, the Master Plan has the following targets:

- The government should ensure that the public and private sectors together invest in ICT research with the aggregate amount equal to or greater than 3 per cent of the total ICT industry value.
- The government should provide a large software development project that requires at least 100 man-years of work, and this project must include research and development activities with the amount of not less than 5000 million baht by 2006.
- By 2004, at least 80 per cent of PC value and 50 per cent of software value consumed within the country should be locally developed.
- By 2004, at least 70 per cent of the Thai software developers should be working in network computing[8] and/or Web services.

On the whole, the past as well as the present IT policies of the Royal Government of Thailand seems to reflect a realistic understanding of the issues being faced by the economy in general and the IT sector in particular. Achieving the targets could take Thailand a long way towards furthering its ICT capability and in harnessing the new technology for the overall development and socio-economic transformation. While the plan is clear about what to achieve, some questions relating to how to achieve the target could be raised. For example, what specific strategies are to be adopted to achieve the target set with respect to software production and exports? What role could trade and investment policies play in achieving these targets? Can ICT production be developed as has happened in the case of electronics? Are there possibilities of collaboration/cooperation with neighboring countries? What is the role of stakeholders other than the private sector, say NGOs, who could play a very constructive role, especially in addressing the issue of intra-national digital divide and harnessing ICT for the rural masses. What is the role of the provincial governments in achieving provincial equity with respect to IT?

Human capital for information technology[9]

Emphasis on human capital formation was nothing specific to IT. During the period of economic crisis, for example, more importance was given to tax incentive schemes for supporting training and skill development in the private enterprises. Although Thailand entered 1990s with a relatively poorly educated workforce, it has been slow at developing policies in this field when compared to neighbouring countries like Malaysia and Singapore. Until 1994 there was no demand side policies to promote training and skill up gradation. In 1994 the Vocational Training Promoting act was introduced, offering tax incentive for company-based training centers. The registered private enterprises could obtain a 50 per cent tax deduction on training expenses in their income tax. In 1995 a new fiscal decree made it possible to deduct 150 per cent of the cost of in-house or outhouse training, when employers trained employees with more than six months of seniority (Department of Skill Development 2000).

Tax incentive to support training through private training institute had, by January 2001, supported 51 companies, which trained 5000–6000 workers on a yearly basis. The other tax incentive scheme with 150 per cent tax deduction had been utilized by about 500 companies during 1990–2000 and there were around 1,90,000 participants per year (Lauridsen 2002). The low utilization of tax exemption scheme has been noted by the Thai Industrial and Innovation survey and by other studies (Arnold *et al.* 2000).

The second major initiative was the skill development fund set up in 1997, wherein soft loans were provided to workers willing to upgrade their skills, knowledge and capabilities. By 2001, whole fund earmarked for the scheme had been utilized. It is found that three quarters of the fund were used by new entrants to the workforce (mostly women) aiming employment in the service sector; a further 5–10 per cent was used by laid-of workers (normally undergoing computer courses) while workers in job used only 10–20 per cent. Another initiative to skill development was the amendment of Skills Development Act, in particular the change in the financing and working of the skills development fund.

Data presented in Table 3.5 tends to suggest that Thailand is likely to achieve the target set out in the Master plan at least with respect to manpower with higher level of qualifications. Yet, the available estimate on the demand for and the supply of IT manpower as well as the projection for the near future as presented in Table 3.5 shows that Thailand will have to live with excess demand situation at least in the near future. During 2002–2006, notwithstanding the near doubling of the supply of IT manpower, the deficit in the year 2006 is likely to be of the order of 26,000. This calls for intensified efforts on the part of government and private sector to address the growing demand for IT manpower.

Given the fact that there is growing demand for computer/software training, a number of private training institutions are operating in this field. The high demand for IT training arises not only from the young generation but also from workers because the private sector in Thailand is in the process of upgrading their operations and expect their employees to acquire IT skills. The Thai Ministry of Education reports

Table 3.5 Comparison between the demand and supply of ICT manpower in Thailand during 1994–2006

Year	Demand (number)	Supply (number)	Balance + = surplus, − = deficit
1994	32,544*	12,501	−20,043
1995	37,217*	14,953	−22,264
1996	42,293*	18,811	−23,482
1997	47,692*	22,126	−25,566
1998	53,517*	24,867	−28,650
1999	59,928*	NA	NA
2000	67,072*	NA	NA
2001	74,640*	NA	NA
2002	92,091*	70,386**	−21,705
2003	106,992	82,986**	−24,006
2004	122,670	97,030**	−25,640
2005	139,154	112,795**	−26,359
2006	156,546	130,502**	−26,044

Notes:
*An average of the estimated demand from two scenarios.
** The supply of workers with a Bachelor's degree and higher.
NA = not available.
Source: Suksiriserekul (2003).

that there were more than 800 computer training schools with approximately 320,000 students in 1999, 11 per cent higher than in 1998. However, from the industrial sector it is discerned that the quality of IT manpower coming out of these institutions is much to be desired. While very high targets have been set by the IT Master Plan, specific strategies towards achieving high quality manpower appears to be necessary.

Here the policy makers may consider the following policy options towards further strengthening the NSI. First, efforts need to be made to foster a strong relationship between academia and industry, involving an active participation by the industry in deciding the curriculum of the university so that the students and faculty get accustomed to a current state of art in the industry. The software park of Thailand indeed makes an earnest attempt in this direction. However, such linkage should perhaps not be confined to units in the park alone. Secondly, it may be advisable to explore the possibilities of fostering relationship with those universities/institutions in neighboring countries known for their IT capabilities. This could be supplemented by creating a more conducive environment for investment, especially in the IT training field. Further, following the strategy that has been adopted by developed countries, relaxing the restriction on the mobility of knowledge workers, may give rich dividends. Given the high variations in the quality of training offered by different training institutions, much could be learned from the DoE accreditation system followed in India.

IV. Trade and investment: Policies, performance and challenges

Among the developing world, Thailand is known for its most liberal policies towards FDI. As early as in the 1970s, while there has been an environment of disenchantment towards FDI among the developing countries, Thailand adopted policies that welcomed foreign investment with minimum of restrictions. The underlying objective of the liberal policies was to reap the benefit from the inflow of capital resource and other intangible assets like technology on the one hand and get better access to world market with the help of multinational corporations on the other. While the policies have changed over the years, in essence it remains as one of the most facilitating policy regime in the region.

Investment promotion in Thailand

The Board of Investment (BOI), set up in 1959 to promote investment, both foreign and local, is the official agency responsible for providing incentives to promote investment in the country. Although investment promotion dates back to more than three decades, the BOI is officially governed by the 1977 Investment Promotion Act, which was amended in 1991 and 2001 to take care of the changing economic conditions. In 1993 the BOI initiated a major shift in emphasis from export orientation to industrial decentralization as a major policy goal.

Guiding principles in investment promotion[10]

As of now the investment policy is governed *inter alia* by two broad objectives: to promote investment in certain priority areas in tune with the national economic development goals and to promote investment in such a way as to promote balanced regional development. In keeping view of the first objective, the BOI is empowered to provide a wide range of fiscal and non-fiscal incentives to investment projects that strengthen the country's industrial and technological base. Hence the BOI promotes investments which:

- strengthen Thailand's industrial and technological capability;
- make use of domestic resources;

- create employment opportunities;
- develop basic and support industries;
- earn foreign exchange;
- contribute to economic growth of regions outside Bangkok;
- develop infrastructure and conserve natural resources and;
- reduce environmental problems.

In view of increasing international competitiveness, every promoted project that has investment capital of 10 million Baht and upwards (excluding cost of land and working capital) must obtain ISO 9000 certification or similar international certification within two years from its start-up date. Special concessions have been offered to promote investment in backward regions with focus on small and medium enterprises. Priority is given to activities in agriculture and agricultural products, projects related to strengthening the national innovation system through technological and human resource development, public utilities and infrastructure, environmental protection and conservation and targeted industries. In general the policy towards investment is highly liberal and transparent and also efficient (Box 3.2), with hardly any performance requirement like export obligation or local content requirement in line with WTO rules.

Box 3.2   Time frame for project consideration and related procedures in Thailand (Days)

Initial project analysis

By the BOI (investment of up to 40 million baht)	60
By the BOI or Sub-Committee (investment of 40–500 million baht)	60
By the BOI (investment of more than 500 million baht)	90

Project modifications

Changes in location, raw material import schedule, registered capital	
Foreign equity share, or reduction in production capacity	5

Additional privileges, product withdrawal, sale or by-products	15
Increase in production capacity, change in type of products, change in method of production, change in exporting condition, or transferring of promoted activities	30

Promotion certificate issuance

Extension of promotion acceptance period	7
Issuing promotion certificate	10

Clearance of machinery imports

Approval of machinery clearance*	3 hours
Approval of use of bank guarantee	3
Approval of bank guarantee withdrawal	7
Approval of master list of machinery	60

Clearance of raw or essential materials imports

Approval of raw or essential materials clearance*	3 hours
Approval of use of bank guarantee	3
Approval of bank guarantee withdrawal	3
Approval of input formula and maximum stocks	30

Granting permission for foreigners

To conduct feasibility studies	15
To work in promoted activities	10
To own land	15
To mortgage land	15
To own a condominium unit	5

* These processes must be done at the Investor Club Association

Source: Board of Investment, 2003, *A Guide to the Board of Investment*, Office of the Board of Investment, Royal Thai Government, Bangkok.

Trade policy regime in Thailand

Thailand has a highly liberalized trade regime that has been particularly facilitating the inflow of raw materials and capital goods for export production. Being a member of WTO, Thai government is complying with WTO Tariff protection commitments. To facilitate trade, in 2000 government announced the new custom valuation system by which there are six methods of custom valuation to estimate the import duty. Government has also made use of IT in the custom departments for the speedy clearance of exports and imports. Licensing requirements has been reduced over the years and today it is required only for 26 categories.

Investment performance

Table 3.6 presents data on the inflow of FDI to Thailand. It is evident that Thailand accounted for about 2.65 per cent of the FDI inflows to developing countries and by 1998, even while the economy was under the grip of financial crisis, FDI peaked at about $7.4 billion and accounting for about 4 per cent of the total FDI inflows to the developing countries. The higher inflow was made possible not only through the liberal trade and investment policies but also because of the abundant supply of cheap labor, political stability, good infrastructure and geographical proximity to investing countries like Japan. Since 1998, there has been a downward trend in inflows and the total inflow in the year 2002 was only about one-sixth of that the country received in 1998. Inflow of FDI since 2002 tends to indicate that a trend reversal is yet to take place. Presumably, in the current context wherein developing countries

Table 3.6   FDI inflows into Thailand in comparison with developing countries and Southeast Asia (US$ million)

Year	Developing Countries	South, East and Southeast Asia	Thailand	Share of Thailand in Developing Countries
1990–95 (Average)	74,288	44,564	1,990	2.68
1996	91,502	56,147	1,964	2.15
1997	193,224	100,067	3,882	2.01
1998	191,284	90,093	7,491	3.92
1999	229,295	105,313	6,091	2.66
2000	246,057	138,698	3,350	1.36
2001	209,431	97,604	3,813	1.82
2002	155,528	86,318	947	0.61
2003	166,337	94,755	1,952	1.17
2004	233,227	137,705	1,064	0.45

Source: UNCTAD, *World Investment Report*, Different years, Geneva, UNCTAD.

in general have opened up their doors for FDI and the MNCs are looking forward to complementary capabilities, the weak Innovation system of Thailand diminishes its attractiveness as an investment location.

The effectiveness of the policy regime in attracting and facilitating investment could be gauged from the fructification rate. During the five years ending 2001, total approved investment was of the order of 29.34 billion Baht and the actual inflow was about 24.6 billion Baht. This makes the fructification rate of nearly 84 per cent, which was one of the highest in the region. As will be seen in the forthcoming chapters, FDI fructification has been much lower in the ASEAN New Comers and therefore much could be learned by these countries from Thai experience.

In terms of the sector-wise distribution of FDI, it is found that electrical and electronics industries accounted for largest share of investment followed by metal products and machinery. During 1997–2001, for example, electrical and electronics industries accounted for nearly 26 per cent of the total FDI inflows into the country followed by metal products and machinery with a share of nearly 20 per cent (Table 3.7).

Investment in ICT and software

Within electrical and electronics industries group, the bulk of the investment has been accounted for by electronics industries (Table 3.8). It is evident that within electrical and electronics industries, electronics accounts for about 90 per cent of the total number of projects approved and more than 95 per cent of total investment. More interestingly, in tune with the policy objectives as well as the vision presented in IT-2010, about 23 per cent of the total number of projects approved in electrical and electronics sector in 2001 has been in the field of software development.

More recent data from the BOI indicates that the cumulative number of promoted companies since 1996 stood at 170 by December 2003. Out of these, 75 were fully owned by Thai companies, 34 were foreign companies and the others were joint ventures. The total investment commitment by 137 companies promoted prior to 2002 amounted to 1632 million Bahts with a mean investment of 11.9 million Bahts. But the actual investment made amounted to only 52 per cent (846.5 million Bahts) with an average investment of 9.7 million Bahts. As we shall see in the next section, the observed rate of investment fructification is much lower as compared to other sectors and for the economy as a whole.

Table 3.7 Sector-wise distribution of foreign investment projects approved in Thailand

	Agricultural products	Mineral and ceramics	Light industries textiles	Metal products and machinery	Electric and electronic products	Chemicals and papers	Services	Total
1997								
Projects	48	22	46	147	97	88	66	514
Share (%)	9.3	4.3	8.9	28.6	18.9	17.1	12.8	100
Investment	7,568	9,240	8,334	28,015	30,123	131,807	85,200	300,469
Share (%)	2.5	3.1	2.8	9.3	10	43.9	28.4	100
1998								
Projects	50	9	76	70	133	66	79	483
Share (%)	10.4	1.9	15.7	14.5	27.5	13.7	16.4	100
Investment	11,777	951	13,250	9,465	59,458	43,427	116,532	254,864
Share (%)	4.6	0.4	5.2	3.7	23.3	17	45.7	100
1999								
Projects	49	9	73	109	143	69	61	513
Share (%)	9.6	1.8	14.2	21.2	27.9	13.5	11.9	100
Investment	11,265	480	8,942	12,452	57,287	41,547	9,423	141,489
Share (%)	8	3	6.3	8.9	14.5	29.4	6.7	100

2000

Projects	72	22	112	195	185	108	67	761
Share (%)	9.5	2.9	14.7	25.6	24.3	14.2	8.8	100
Investment	23,127	9,991	23,937	26,122	71,613	54,449	3,407	212,649
Share (%)	10.9	4.7	11.3	12.3	33.7	25.6	1.6	100

2001

Projects	46	13	57	135	173	84	67	575
Share (%)	8	2.3	9.9	23.5	13.5	14.6	11.7	100
Investment	15,273	5,411	12,150	25,374	51,855	69,908	29,649	209,623
Share (%)	7.3	2.6	5.8	12.1	24.7	33.3	14.1	100

1997–2001

Projects	265.00	75.00	364.00	548.90	731.00	415.00	340.00	2,846.00
Share (%)	9.31	2.64	12.79	19.29	25.69	14.58	11.95	100.00
Investment	69,010.00	26,073.00	66,613.00	101,428.00	270,336.00	341,138.00	244,211.00	1,119,094.00
Share (%)	6.17	2.33	5.95	9.06	24.16	30.48	21.82	100.00

Note: Projects are given in number and investment figures are given in million baht.
Source: Board of Investment (2003).

Table 3.8 Distribution of investment across different projects in electrical and electronic products in Thailand (2001)

Industries	No. of projects	Investment (million Baht)	Employment	
			Thai	Foreign
Electronic Industry and Electric Electrical Appliance	173	51,855	31,753	876
Manufacture of electrical products	5	888	1,677	62
Manufacture of parts or equipment used for electrical	5	395	381	13
Manufacture of electric lamps	1	41	60	1
Manufacture of insulated wires or cables	2	80	27	1
Manufacture of parts or equipment for other appliances	3	1,520	474	22
Manufacture of electronic products	5	1,034	1,845	15
Manufacturing of office equipment computing or accounting	1	115	425	5
Manufacture of household appliances	1	2	115	–
Manufacture of radio, television or telecommunications	3	1,199	995	17
Manufacture of professional and scientific measuring	1	39	123	1
Manufacture of parts or supplies used for electronic	65	38,991	19,262	362
Magnetic components, including telescopic antenna	5	301	634	91
Connectors	2	71	162	5
Printed circuits boards	2	658	255	10
Plugs and sockets	1	55	220	10
Acoustic parts, including microphones, ear-phones	1	105	196	10
Micro-motors	2	163	392	9
Diodes	1	498	95	–
Computer components, including storage equipment	6	701	1383	17
Electronic sub-assemblies, including PCBs	7	839	526	17
Flat, shielded, coaxial or signal cables	3	436	517	10
ICs	3	2,304	345	6
Capacitors	3	500	86	–
Relays	1	109	28	2
Switches and keyboards	1	5	19	1
Embedded system design	1	3	21	1
Software	40	764	1,430	176
E-commerce business	4	32	70	12

Note: Investments are given in million baht.
Source: BOI.

In case of fully Thai-owned companies the total investment commitments amounted to 532 million Bahts and the actual investment was of the order of 389 million Bahts (73 per cent). When it comes to fully owned foreign companies, total investment commitments were of the order of 113 million Bahts and the actual investment was about 90.8 million Bahts (80 per cent). In the case of 55 joint ventures (1996–2002), total investment commitment was of the order of 986 million Bahts with an actual investment of 365.9 million Bahts (35.7 per cent). Thus it appears that investment realization has been the highest in the case of fully owned foreign companies followed by Thai companies and joint ventures. It is beyond the scope of the present study to analyze how to account for the observed differences in the rate of investment realization across different ownership categories. Further enquiries in this direction may lead to important insights.

Total employment commitment by the 170 promoted companies has been of the order of 4207, whereas the actual employment generated by 2003 has been only of the order of 1969. Out of the realized employment, 1184 (60 per cent) were accounted by Thai firms, 165 (8 per cent) by foreign firms and 620 (32 per cent) by joint ventures. Of the foreign firms, 25 per cent are from Japan, 13 per cent from the US and 9 per cent from the UK. Other countries with foreign investment in the software sector include Canada (7 per cent), France (7 per cent), Singapore (7 per cent) and the Taiwan Province of China (5 per cent).

Table 3.8 also presents data on employment potential with breakup of foreign and local. It is evident that the number of foreign employment per million Baht of investment in the electrical and electronics sector is found to be 0.016. But when it comes to software, the corresponding ratio is 0.12, more than seven times the industry average.

V. Present state of ICT use

To reflect on the present state of IT use, we shall examine some of the indicators like the use of telephones, both fixed and mobile, Internet and computers. In addition we shall also reflect on the use of ICT in government and other sectors of the economy.

Telecommunication

Since the telecommunication sector of Thailand has been subjected to detailed enquiry, the present study do not intend to get into many of the details, which are available in earlier studies (ITU 2002b, Tangkitvanich and Ratananarumitsorn 2002, Cairns and Nikomborirak 1997). We shall begin with a profile of telecommunication sector in terms of certain broad indicators, in comparison with other countries taken up for analysis in this study. Table 3.9 shows that in terms of most of the indicators of telecom development Thailand holds a position much above other countries. In the case of fixed lines, which are provided by TOT,[11] Telecom Asia,[12] TT&T[13] and CAT, the density increased more than six fold during the last decade. The waiting time has considerably declined from about five years in the mid-1980s to about a little more than a year in 2002. The number of public telephones in the country recorded an Annual Compound Growth Rate (ACGR) of nearly 16 per cent during the last decade. By March 2003 there were about 13.1 telephones per 100 people. There are many other dimensions of telecom growth and penetration that Thailand could be proud of.

Table 3.9 Indicators of telecommunication development in Thailand in a comparative perspective (2002)

Telecom indicators	Cambodia	Lao PDR	Myanmar	Vietnam	Thailand	Lower middle-income countries
Lines per 1000 people	3	11	7	48	105	164
Lines per 1000 people in the largest city	19	65	32		452	524
Waiting list (000)		5.9	93.5		710.2	
Waiting time		1.1	5.3		1.6	1.9
Lines per employee	61	45	43	49	222	110
Revenue per line ($)	705	437	61	359	637	637
Cost of local call ($ per 3 minutes)	0.03	0.02	0.05	0.02	0.07	0.04
Mobile per 1000 people	28	10	1	23	260	99
Outgoing traffic (minutes per subscriber)	278	138	27	17	52	110
Cost of call to US ($ per 3 minutes		6.37	0.36		1.54	2.09

Source: World Bank (2004) and World Development Indicators (2004).

While the inter-temporal growth has been remarkable, the question remains whether the country has been able to fully exploit its potential? To answer this question, it is helpful to place the performance in a comparative perspective. Here we shall compare Thai performance with the average of lower middle-income countries and also with neighboring Vietnam. Despite the fact that Thai's per capita income is almost five times higher than Vietnam, the number of telephone lines per thousand people in Thailand is only a little more than twice that of Vietnam. The telephone penetration in Thailand is also low compared with the lower middle-income countries.

In developing countries the intra-national digital divide – disparity in the access to IT infrastructure across different sections and regions within the country – has been found to be as pronounced as the international digital divide. How does Thailand perform in this respect, which has been highlighted as a major task to be addressed by IT-2000? Table 3.9 reveals that the number of telephone lines per 1000 people in the largest city of Thailand is almost 4.5 times higher than the average. In case of all lower middle-income countries considered, the ratio is lower (3.1). As we shall see later in this book, in case of Vietnam, it has been observed that during 1995–2000 the telephone lines recorded an ACGR of over 31 per cent, which was almost equally distributed across different regions in the country.[14] In Thailand, the metropolitan areas, particularly the capital city and the surrounding areas, were having a teledensity of 54 per cent, whereas in other 76 provinces the teledensity was as low as 6.1 per cent (Tipton 2002).

It appears that more effort is called for to ensure that telecom services in Thailand is offered at competitive prices. Fixed-line telecommunication tariff in Thailand remained constant (connection charge of 3350 Baht and monthly rent of 100 Baht) for a number of years. There is no differential pricing for business and residential connection. While a local call costs only $0.02 in Vietnam and Lao PDR, and $0.04 in the lower middle-income countries, in Thailand it costs ($0.07) more than three times that of Vietnam and almost twice that in lower middle-income countries. However, one must add that in Thailand the recorded cost is for unlimited time whereas in other countries like Vietnam it is for three minutes. Such pricing strategy reflects on the revenue per line. While the cost of call is more than three times higher in Thailand as compared to Vietnam, the revenue per line in Thailand is only 1.7 times higher. While the cost of local calls is not strictly comparable with other countries, Tangkitvanich and Ratananarumitsorn (2002) show that Thai consumers are made to pay more than their counterparts in other countries when it comes to domestic long-distance calls and international calls.

In the case of mobile telephones, significant progress has been made in terms of growth and diffusion. There are seven mobile service providers and of them Advanced Information systems (AIS) is the market leader. As of May 2003 it was estimated that mobile penetration was approximately 30 per cent. However, it has also been argued that the industry is not highly competitive and prices are shown to be higher as compared to neighboring countries (Tangkitvanich and Ratananarumitsorn 2002).

The performance of the Telecom sector (both mobile and fixed) has to be seen in the context of telecom market structure that evolved over the years and that has been strongly influenced by policies initiated from time to time. Till recently, the telecom market in Thailand has been the monopoly of the state. The Telephone Organization Thailand (TOT) has been the only operator for domestic (local and long distance) market including neighboring countries with common borders. The Communication Authority of Thailand (CAT) had the complete monopoly in terms of international services including Internet. Both of them, however, were entitled to provide other related telecom services like paging, cellular and Very Small Aperture Terminals (VSAT).

In the context of limits set by the technical, financial and other organizational matters for providing technical services on the one hand and growing demand[15] for telecom services on the other, certain policy reforms were introduced in the mid-1980s. This took the form of government relaxing its monopoly control in the telecom market and permitting private sector participation. However, the private sector participation has been visualized in the form of Build–Transfer–Operate Scheme (BTO).[16] The introduction of BTO has to be seen as a means of overcoming the complicated process of amending a number of telecom laws. The BTO has led to the entry of a number of private sector firms into the telecom sector. As of 2001 there were 30 concessionaries that obtained concessions from TOT and CAT at varying terms and conditions.[17]

In 2003 the government has brought about major changes in the telecom sector, which involved the privatization of both TOT and CAT (now called TOT Corporation Plc and Thai Post Co Ltd respectively). This is expected to create a more competitive environment and hence motivate the existing operators to prepare for further competition once the telecom liberalization as per WTO comes into full effect in 2006. The government is also in the process of establishing two regulatory authorities, National Telecommunication Commission and National

Broadcasting Commission, in accordance with the Frequency Allocation Act. Another issue being currently tackled relates to BTO concession conversion. Since an exploration into this and other related issues is beyond the scope of the present study, we only reiterate the need for creating a more competitive environment such that the country is able to fully utilize its potential.

Internet

The Internet was introduced to Thailand in 1991 through academic and research applications. The first Internet in Thailand was the Thai Social/Scientific, Academic and Research Network (ThaiSarn). Starting from only 9600 bps international link in 1992, ThaiSarn was matured within about three years of its introduction with the first 2 Mbps international link in September 1995. ThaiSarn became the main academic and research network of Thailand with a number of information servers providing document archives, freeware/shareware mirrored archives and major local information such as the Golden Jubilee Network, which hosts a wealth of information about Thailand in Thai language. The collection of servers is called "PubNet". In November 1997, ThaiSarn launched the Public Internet Exchange (PIE) to supplement the PubNet project. PIE allows local commercial ISPs (through their own investment and connection license) to exchange domestic traffic without leaving Thailand. The project was so successful that after one year of experiment, the participants of PIE decided to provide funding to sustain the project (Koanantakool 2001).

In 2003 there were 18 ISPs, of which the top five accounted for over 70 per cent of the total traffic. These five also accounted for majority of the subscribers – both dial-up and leased line. CAT issues ISP licenses because it is having the monopoly over international communication. ISPs are expected to provide 32 per cent of their shares to CAT. Limits on foreign ownership in ISPs were raised to 49 from 20 per cent in 1998. Since the local call charges are fixed regardless of the duration of call, demand is tilted in favor of dial up and the demand for broadband is limited. The dial-up Internet access price has two components: the ISP charge and the telephone charge. The present pricing of telecommunication acts as a factor favoring Internet use. Since the price of local call is fixed regardless of the time used, Thailand has one of the lowest dial-up Internet prices in Southeast Asia (ITU 2002b). There have been attempts to do away with the ISP charge completely for dial-up Internet access. Though this has not been realized, one of the ISPs, Telecom Asia, offered promotions for its fixed-line customers by providing free Internet access through its ClickTA ISP.

As a result of the series of initiatives by the state, there has been significant progress in the use of Internet in the country. By 2001 there were about 3.5 million Internet users in the country, which accounts for about 5.6 per cent of the population. Yet English remains a major hurdle for large sections of society in accessing Internet. It has also been observed that there is gender equality regarding Internet use – almost 50 per cent of the users are females. As envisaged in the policy, dial-up Internet is available in all the provinces at the cost of a local call. At the same time, it needs to be noted that, as in many other developing countries, the capital city accounts for about 71 per cent of the users and about 90 per cent of the users are in the urban centers though the urban centers accounts for only about 30 per cent of the total population.

IT in government

Various projects were undertaken to promote the use of ICT in government at different levels. The projects include the Government Information Network (GINET), Chief Information Officer (CIO) program and the Chief Executive Officers Program (CEO) program.

GINet

There are two major components to the core IT infrastructure of government: the GINet and the common information services to all ministries. The Government Information Technology Service Program at NECTEC manages both functions. GINet is the government networking backbone which links every province in Thailand with high-speed communication lines at the speed of 155 Mbps. High-speed access networks are to be made available in all 1000 districts through leased circuits, ISDN or ADSL technologies. GINet consists of a nationwide Asynchronous Transfer Mode (ATM) network running on TOT's existing and new optical fiber network. By April 2000, about 20 provinces were having access to GINet backbone. Various initiatives have also been made to make use of IT in the provincial administration.

The CIO program

Started in 1998, the program aimed at appointing a CIO in every ministry, department and state enterprise to oversee the unified IT development plans at both the departmental level and the ministerial level. The responsibility of a CIO included drafting of the organization's IT

Master Plan and transforming relevant national IT policies into organizational actions. NECTEC and the Office of the Civil Servant Commission jointly provided the CIOs with required training. The government CIOs are instrumental in ensuring smoother information flow across ministries, more efficient information sharing and improved decision-support system for the country through the use of GINet and a common set of specifications to allow the private sector to carry out IT projects for the government more efficiently.

A related project aims at increasing IT awareness at the highest government level. The CEO approved by the cabinet in 2000 calls for two high-ranking officers (permanent secretary and director general) within each government organization to attend half-a-day training session on the benefits of ICT. By the end of 2001, all the CEOs have attended the course.

e-commerce

The importance of electronic commerce (e-commerce) has been recognized by NECTEC and NITC since 1992, when NITC set up a subcommittee on Entrepreneurship Development Institute (EDI) for international trade. Since then, NECTEC has been developing EDI service organization with the Customs Department, Thai Airways International, CAT, TOT, the Federation of Thai Industry, the Chamber of Commerce, Association of Freight Forwarders, and so on. In 1998 a joint venture company called TradeSiam was set up with the private sector as the majority shareholders and the government as the minority. The EDI subcommittee was subsequently renamed "Thailand EDI Council" (TEDIC). The NITC assigned NECTEC to develop an electronic commerce framework to develop recommendations on the roles and responsibilities of government agencies. One of the objectives of the plan is to facilitate private sector involvement in evolving domestic and international e-commerce. In January 1999, the cabinet approved a proposal by the Ministry of Science, Technology and Environment to set up the Electronic Commerce Resource Center (ECRC) to ensure the smooth development of e-commerce in Thailand through awareness creation, training program and by setting up an information center. In 2000 government approved the National Policy Framework for Electronic Commerce drafted by ECRC. Electronic Commerce revenue in the country was estimated at US$90 million in 2000 and was expected to reach $2.3 billion by 2004. Yet it has been argued that Thailand lags behind the neighboring countries in terms of e-commerce (ITU 2002b).

TradeSiam – Thailand's national EDI provider

The TEDIC, one of the subcommittees under the NITC, proposed the creation of TradeSiam as a joint venture company between Thai government agencies and the private sector mainly to facilitate international trade. It started a limited pilot service in December 1998 and became fully operational in 1999. TradeSiam serves as a center to provide EDI services between government agencies and the private sector. In order to operate efficiently, TradeSiam is managed as a private company where it positions itself as a national EDI service provider.

IT in education: SchoolNet

The SchoolNet Project initiated by NECTEC in cooperation with CAT, TOT and the Ministry of Education was started in 1996. The project has the following specific objectives:

- provision of Internet connection and technical support to schools;
- promotion of content development and training the teachers;
- promoting the use of Internet in classroom activities.

During the first year of implementation, 20 schools were connected and in 1999 the Cabinet approved the expansion of the project to cover 5000 schools across the country. By 2003 there were 4751 schools in the SchoolNetnetwork. The SchoolNet Web site[18] serves as the information center for teachers and students and is the portal for school Web sites in all regions. Some 1289 schools participate in the activities of the Web site. By 2003, total number of 505,120 teachers had undergone training in ICT.

The Ministry of University Affairs is the lead organization in the development of Inter-University Network (UniNet) in accordance with a Cabinet Resolution in October 1997. Its mission is to develop a high-speed information highway and establish a distant learning network for the university system. By 2003, 30 institutions had joined together through the UniNet fiber-optic network. The project has four elements:

1. join all universities and colleges in the network;
2. develop self-access learning centers within universities, including the establishment of electronic library facilities in Campus Networks;
3. develop courseware for joint use by university members as well as information databases and instruction via video conferencing; and
4. capacity building and personnel development to enable research, management and application of new learning technologies.

In 2001, the Cabinet approved the creation of a National Education Network Project, or EdNet, which merged the UniNet and SchoolNet networks into a National Education Network, with the Ministry of University Affairs charged with the development and management of the infrastructure. The Ministry of Education is responsible for installation of equipment and computers in the schools and learning centers in each province. A committee has been appointed to review the design of the ICT system and plan the expansion of the network to include educational institutions from the basic through the tertiary levels.

The use of IT in Thailand is not confined to government and education alone. Significant use of IT is being made in the field of health care, agriculture[19] and other activities related to poverty reduction. However, a detailed enquiry into the extent of use and its impact on all the sectors is beyond the scope of the present study.[20]

Private sector

Private sector in Thailand has been in the forefront with respect to the use of IT. By 2003 about 6460 e-commerce Web sites were established by Thai private sector companies. E-commerce sites are growing in different sectors like tourism, computer and Internet, apparel, cosmetics, handicrafts, jewelery, restaurant and so on (UNESCAP 2002). It is understood that small and medium enterprises are also increasingly having their own web sites and engage in e-commerce.

VI. Concluding observations

Among the developing countries, Thailand has a longer history of liberal trade and investment policies which in turn was instrumental in bringing about higher output growth along with structural transformation wherein the industrial sector, and more specifically the manufacturing sector, emerged as the most vibrant sector of the economy. The liberal trade and investment regime coupled with other facilitating environment like good infrastructure and abundant supply of cheap labor led to substantial investment in the field of IT hardware production. Thus, over the years electronics has emerged as one of the major sources of employment and export earning. Today almost all the world leaders have their presence in Thai electronic industry. However, the electronics industry in Thailand has been characterized by sticky specialization in a few low technology products leading to low value addition, poor forward and backward linkages and high import intensity and getting locked up in the low end of the electronics value chain.

There is some merit in the argument that the emergence of a lopsided production structure has been an outcome of the policies followed hitherto towards investment. While encouraging investment, the incentive structure was not tuned to induce the companies, both foreign and local, to invest in skill upgradation and knowledge generation. More importantly, unlike India, there has not been any significant attempt towards fostering a vibrant national system of innovation linking different actors involved in knowledge generation and diffusion. This has led to limited skill and knowledge base, which in turn acted as a stumbling block for the establishment of high value-adding skill-intensive activities like IT software and services.

The single most important hurdle that Thailand faces in making headway in the sphere of IT is the scarcity of human capital, in terms of both quantity and quality. Despite various initiatives, especially during the post-crisis period, the IT sector of Thailand is faced with an excess demand situation. While human capital cannot be built overnight, the policy makers may consider a two-pronged action towards increasing the quantity and quality of IT manpower. More targeted policies for attracting investment into the IT manpower training may be beneficial. In the short run relaxing the restriction on the mobility of IT manpower may give rich dividends. Toward improving the quality of manpower, it is necessary to foster strong relationship between academia and industry by involving an active participation by the industry in deciding the curriculum of the university so that the students and faculty get accustomed to the current state of art in the industry. While the software park of Thailand makes an earnest attempt in this direction, there is a need to scale these activities up. It may also be advisable to explore the possibilities of fostering relationship with those universities/institutions in the neighboring countries known for their IT capabilities. Given the high variations in the quality of training offered by different training institutions, much could be learned from the accreditation system followed in India.

Realizing the importance of IT, the Thai government initiated certain pioneering efforts in the form of new policies and institutional structures which culminated in the formation of a separate ministry for IT not only towards developing an ICT base in the country but also in the widespread use of technology in different sectors of the economy including government. While the first IT policy, IT-2000, aimed at laying foundations for development and use of new technology, the IT-2010 provided a long-term vision with focus to further strengthening the innovation system and also to facilitate the transition of Thailand into a knowledge-based economy and society, wherein creation, collection, dissemination and utilization of knowledge emerge as major instruments of economic and social development. The vision got translated in the IT Master Plan (2002–2006) that has highly ambitious targets in terms of ICT production and use as manifested in e-governance, e-commerce, e-society, and so on. These policies indicate a realistic understanding of the issues being confronted in terms of developing and harnessing the new technology for development. Yet the overall approach of the policy appears to be one of highly centralized decision making with limited role for the provincial authorities. Also, the role that Civil Society Organizations could play in achieving the targets set by the government needs to be further explored.

However, there has been attempts in the recent years towards building up an innovation system as is evident from institutional interventions and policy measures for strengthening skill base, promoting R&D investment and fostering an interface between academia and industry. Also, there is evidence to suggest a greater orientation in the private and public sectors towards innovative activities. Moreover, the government has undertaken initiatives to foster relationship with countries outside ASEAN like India that may lead to beneficial cooperation in the field of skill-intensive areas like IT and software. On the whole, with the marked revival in the economy during the recent years and a series of initiatives to strengthen the innovation system along with the liberal trade and investment regime followed over the years, the future appears to be more promising than the past.

4
Cambodia: Between Pentium and Penicillin?

I. Introduction

Cambodia is, perhaps, one of the latest entrants to the club of developing countries that shifted from the import-substituting growth strategy and embraced the outward-oriented growth path. But, as compared to other countries, the task that Cambodia had to undertake has been undoubtedly more arduous because the reformers had to inherit a devastated and destabilized economy.[1] The rest of the world, however, has been highly sympathetic to the cause of Cambodia as evident from the formation of International Committee for Reconstruction of Cambodia (ICORC), through which several governments pledged around $2.3 billion for the period 1992–96. Bretton Woods Institutions also helped in macroeconomic stabilization followed by a package of structural adjustment beginning from 1994 (Kannan 1997). Moreover, Cambodia emerged as a major action point for most of the leading NGOs in the world. As a result, notwithstanding the difficult task at hand and the rocky road that the country had to traverse, the achievements during the last decade appear remarkable.

The industrial sector, though comprising mainly of traditional industries like garments and leather goods and dominated by small enterprises mostly under single proprietorship (55 per cent), recorded an ACGR of about 14 per cent during 1993–2000, and the pace of growth continued with 12.9 per cent in 2001 and 17.7 per cent in 2002. As a result, the share of industrial sector in GDP increased from about 13 per cent in 1993 to over 30 per cent in 2003. The primary sector, mainly crop production, has been less dynamic with a recorded growth rate of only 3.8 per cent during 1993–2000 and followed by negligible or negative growth in 2001 and 2002 and accounted for about 35 per cent of the GDP in 2002. While the service sector in general was stagnant with the lowest growth rate of only 1.8 per cent during 1993–2000, it has shown a revival with over 4 per cent growth since 2000. Within services, tourism emerged as a major growth sector recording an ACGR of about 9 per cent. In the external sector, there has been improvement in the trade performance as manifested in the reduction in trade deficit, though the economy depends on official transfers to finance the current account deficit.

These achievements notwithstanding, there is the need for further improvements in growth dynamism of different sectors of the economy and enhancing governance in general (Kato *et al.* 2000). Given the fact that large area of land remains to be brought under cultivation and that there is scope for improving the yield levels of almost all the crops, significant improvements are called for in the agricultural sector so that it contributes towards the welfare of the rural masses. The industrial sector also calls for major structural transformation such that a more diversified and modern industrial sector is established. In the case of infrastructure, there is the need for attracting further investment into the area of power generation (with only about 8 per cent of the households having access to electricity), road and rail transport and so on.[2] When it comes to the social sector, there are more challenges: with a population growth rate of about 2.4 per cent and almost 55 per cent of the population being under the age of 20 and with a life expectancy of about 56 years, Cambodia has to find substantial resources and create institutional structures for enhanced social sector performance in general and for providing health and education to its growing population in particular.

Concomitant with the concerted efforts to develop a market-oriented economy, there have also been a number of initiatives to harness the new technology for development. As early as in 1993, Cambodia adopted the mobile communication technology to address the telecommunication needs of the rural people and it became the first country in the world to have more mobile telephones than fixed telephones. In 2000 the government set up the National Information Technology Development Authority (NIDA) directly under the prime minister of the country with a view to develop the IT sector and harness the new technology for the development of the country.

Against this background this chapter begins with an examination of the present state of Innovation system in the IT sector (Section II). As in the case of other country case studies that follow, the third section undertakes an examination of trade and investment regime and proceeds to analyze the present state of ICT production and use in the country (Section IV). The final section sums up the discussion and presents the concluding observations.

II. Innovation system in the IT sector

The policy framework

Today, almost all the developing countries in the world have either developed or are in the process of developing a policy framework that governs the ways and means by which ICT is harnessed for addressing different development issues. Given the fact that countries vary not only in terms of development needs but also in terms of their capability to use new technology, the form and content of the policies is found to be varying from one country to another. One of the distinguishing characteristics relates to the emphasis that they give to IT production versus IT use, while the other relates to the role of the state versus the market. It is often held that the job cannot be left entirely to the market, as the market is unlikely to be able to guarantee that the investment in information for citizens will meet all of society's needs (Oranger 2001). While countries, which are endowed with an advanced innovation system focus on both use and production of IT, countries with a weak innovation system focus more on IT use with the implicit assumption that IT production is highly skill, capital and technology intensive and that it is beyond the reach of less developed countries. But, one of the arguments that the present study tried to develop (see Chapter 1) has been that the characteristics of IT goods and services production are such that it is not impossible for many developing countries to participate in the global production process of ICT if appropriate trade and investment policies are in place along with an innovation system.

Given the state of under development of the economy and the imperative to address the basic needs of the people, an innovation system in its full-fledged form is yet to evolve in Cambodia. Yet there have been a number of policy initiatives and institutional interventions undertaken in the recent past, which has set the beginning of the making of an innovation system. Cambodia is yet to finalize a policy towards IT though a draft ICT policy has been prepared by late 2004. The process was launched in July 2003 at the National Summit on ICT policy and strategy,[3] convened under the chairmanship of the prime minister. Subsequently, five different working groups were appointed with the task of preparing ICT policy for the country. The draft ICT policy deals with the following broad issues like the national commitments, regulatory framework, human capacity, content development, IT infrastructure and IT enterprise development.[4]

The policy upholds the government's commitment towards harnessing ICT at different levels for addressing development issues like poverty, illiteracy and disease. In this process the policy envisages to actively foster and enhance bilateral, regional and international cooperation. The importance attached to the new technology is evident from it being placed directly under the prime minister. The draft policy also addresses the need to develop human resources for ICT and calls for standardization of curriculum in IT education system throughout the country. The policy prescription on IT infrastructure and encouragement given for investment in ICT sector for promoting its production in the country reflects the understanding of the policy makers of the present state of IT infrastructure and ICT production.

On the whole, the two distinguishing characteristics of the draft Cambodian IT policy are the emphasis on IT use and its production with a greater role for private sector. Given the present state of domestic capital and the low rate of savings coupled with limited technological capability, the focus on private sector, especially the FDI, also has to be seen as an initiative in the right direction. The draft policy, while giving importance to IT use, also underlined the need for developing complementarities for promoting IT use like the local content development by taking into account the specific development requirements of Cambodia. Yet, in developing the needed local content and empowering the people to make effective use of the same, there is the need for strengthening the innovation system, bringing together different stakeholders including the private sector, the academia and the Civil Society Organizations. Studies, however, tend to indicate that at present productive interaction between the government and non-government sectors, in particular the Civil Society and the private sector, is limited (Kato *et al.* 2000). While the emphasis on the role of private sector is in the right direction, one should not forget the fact that the private sector investments are guided by the market test of profitability and many of the ICT projects, especially those addressing the needs of rural poor, might not be profitable, at least in the short run, and hence may not be undertaken by the private sector at all. Hence, there might be the need for public–private–NGO partnerships that could be instrumental in harnessing ICT for development.

IT infrastructure: Radio and television

In a low-income economy like Cambodia wherein the population is more sparsely settled in a number of provinces, the importance of the means of mass communication like radio and television, as instruments of development communication, cannot be overemphasized. However, given the low per capita income, the use of both radio and television in the country is at a very low level. The number of radios per 1000 population in the country was only 105 in 1995 and during the six years that followed (in 2001) only a marginal increase (119/1000) was witnessed in the use of radios. This compares poorly with 156 for the low-income countries in general and 287 for the East Asia and Pacific. When it comes to television receivers that are costly and perhaps beyond the reach of many, the rate of use is found to be much lower and, more importantly, there has been no increase in the rate of use during the six years under consideration. To be more specific, the number of television sets per 1000 people in Cambodia was only 8 in 1995 and it remained at the same level in 2002.[5] Notably, in 2002, television penetration in the East Asia and the Pacific was as high as 317, and 91 in the case of low-income countries.

Fixed telephone

The MPTC is in charge of all aspects relating to telecommunications in the country. The ministry not only regulates the behavior of other actors by being the policy maker, but is also an active participant in the provision of telecom services in the country either directly as a service provider or indirectly as a partner of other telecom companies.

Unlike many other countries where the landline is generally under the state monopoly, in Cambodia, there are two other private companies involved in the provision of fixed lines. The first one being Camintel, which is a joint venture between PT Indosat of Indonesia (49 per cent) and the Kingdom of Cambodia, was established in May 1995. Its telecommunication infrastructure was taken over from the United Nations Transnational Authority in Cambodia (UNTAC). As soon as Camintel took over, it upgraded and expanded the system. Camintel officially started operating on February 26, 1996, providing its customers with a reliable telecommunication network nationwide. Camintel has the highest network coverage in Cambodia, covering all the provinces except five, namely Kanda, Krong Paili, Otdar Mean Chey, Neak Loeng and Bavet.

The second operator, Camshin (Cambodia Shinawatra Co. Ltd), was established in 1993. Camshin is a subsidiary of Shin Satellite PLC, Thailand. Shin Satellite PLC is one of the leading satellite operators in Asia and is backed by the leading telecommunications group Shin Corporation. Shin Satellite PLC owns and operates four communications satellites, namely Thaicom 1, Thaicom 2, Thaicom 3, and the iPSTAR. In the beginning, its operations were confined only to providing fixed phone lines in Phnom Penh. Since then Camshin has developed and expanded its services considerably. Camshin now provides both fixed and mobile phone services (GSM 1800 and GSM 900) in the whole of Cambodia.

New technology: Mobile telephone

Cambodia presents a typical case where the specific technological characteristics of mobile telephone technology make it possible to provide telecom access to people with low income levels and remotely and sparsely settled with relatively low investment. It may not be an exaggeration to say that had there been no mobile technology, large majority of the Cambodians would not have had any access to telecommunication facility at all. It is also incidental that the emergence of mobile technology also coincided with Cambodia's efforts to reconstruct its infrastructure including telecommunication facilities. Moreover, the government, which has been constrained by its ability to invest huge resources to provide fixed telephone, has been highly proactive in terms of allowing private sector to enter the mobile sector and to fill the vacuum at least partly. Hence while in many countries most of the mobile phones act as a complement to the landlines, in Cambodia it emerged as an alternative to land lines. It is therefore no wonder that Cambodia has the distinction of being the first country in the world to have more mobile subscribers than the fixed telephone subscribers as early as in 1993 and by 2000 more than four out of five telephone subscribers were using wireless phones (ITU 2002c).[6]

There are at present four players in the mobile market (Cam GSM, Camshin, Cascom and Camtel). Cam GSM is a joint venture between Luxemburg-based international Cellular company, Millicom (58.4 per cent), and the Royal Group of Cambodia. Cam GSM has been the largest mobile operator in Cambodia since 1998, just one year after launching its GSM network. While it is claimed that its mobile network covers all the provinces, the majority of its customers are in Phnom Penh. Ever since its operations began in 1997, a total investment of nearly $100 million has been made in Cambodia by mid-2001. A new project with total investment of $18 million to build cellular transmission stations and antennae in 45 additional districts throughout Cambodia has been implemented in 2002. Tele2, a sister company of Cam GSM, obtained a license for international gateway which it launched in November 2000. In addition, Tele2 launched a broadband fixed wireless access service in March 2001 (ITU 2002c).

Cambodia Samart Communications Company Ltd (CasaCom) is a joint venture between the Samart Group of Thailand, Telekom Malaysia and the Government of Cambodia (30 per cent). With an initial investment of about 7.5 million, its operations began with an analogue network in 1992 and GSM network in 1999. Camshin (Cambodia Shinawatra) was originally a joint venture with the government to provide a Wireless Local Loop (WILL) network. In 1997, Camshin was converted into a fully owned subsidiary of Shinawatra Group. It was granted a GSM license till 2032 and the 1800-frequency network was launched in 1998. Camtel (Cambodia Mobile Telephone Company) owned by the CP group of Thailand was the first to enter the Cambodia's mobile market in 1992. Another company, Tricelcam, which was a joint venture between TRI of Malaysia and Cambodian Government, ceased its operations in 1998. At the time of closing down, the company had about 3500 subscribers.

Internet and computers

Though the beginning of e-mail in Cambodia dates back to 1993, Internet with commercial services was launched only in 1997. Until mid-2001 there were only two ISPs. The first one was the state-owned Camnet and the second one being Bigpond. While Camnet is a part of MPTC, it is supposed to be run as a commercial entity. Given the constraints linked to its government ownership, the ISP has been often slow to respond to the market needs because changes must run through the regular decision-making process within the government. Needless to say, being a government entity, Camnet is in a disadvantageous position as compared to its private competitor (ITU 2002c).

The second operator, Bigpond, launched its services in 1997 when the MPTC and Telestra agreed on a duopoly until 2002, providing Bigpond with protection from potential entrants and determining revenue sharing with MPTC. The share had increased to 40 per cent in 2001 but dropped to 20 per cent when Bigpond agreed to an early ending of duopoly and opening the market for others by the middle of 2001. This has led to the entry of Mobitel, the leading player in Mobile telephone, and Camintel[7] into the Internet market. In March 2001, Mobitel launched a broadband wireless service covering an area of 14 kilometers in Phnom Penh. It was stated that by December 2001 it had some 800 subscribers, mainly small and medium enterprises, some large corporations and some individuals. Within the short period it has brought about innovations like launching of a bilingual portal (English/Khmer) and now provides its customers with a free e-mail account and facilitates sending and receiving e-mail in Khmer. It has also introduced pre-paid Internet cards.

The Open Forum of Cambodia, a leading NGO, has been providing e-mail services since 1994, almost two years before the two commercial ISPs started their operations. Yet it is understood that the MPTC denied the request of Open Forum to be an ISP (Ratanak 2001).

Human capital

The higher education system in the country comprises of five public universities, three semi-independent specialized institutes of faculties and six recognized private higher education institutes. It is estimated that all these institutes together turn out a total of 25,000 students. It is observed that as in the case of general education, the per cent of women in the total number of students is significantly lower (Table 4.1). Out of the higher education institutions listed in the table, only two – Royal University of Phnom Penh and Norton University – offer degree

Table 4.1 Higher education enrolments in Cambodia (2001)

Institutions	Total students	Women(%)	With scholarships (%)
Public			
Royal University of Agriculture	815	12	91
Royal University of Fine Arts	519	23	100
Royal University of Phnom Penh	4,705	26	53
Maharishi Vedic University	423	6	86
National Institute of Management	8,526	36	8
Institute of Technology of Cambodia	268	6	77
Faculty of Law and Economic Science	3,196	23	25
University of Health Sciences	852	26	100
Private			
Norton University	3,619	23	
Institute of Technology and management	543	26	
Faculty of Management and Law	612	22	
Faculty of Washington DC	281	32	
Institute of Management and Economics			
International Institute of Cambodia	329	28	
Total	24,982		

Source: Louise and Frances (2002).

programs in IT-related fields. The outturn of IT graduates from these institutes is estimated at about 200–300 per year. CISCO, at the instance of NIDA, also started imparting IT training. Also a number of NGOs are found involved in capacity building and IT training along with a large number of private training centers offering short-term courses in IT. Thus there are multiple actors involved in the generation of human capital for IT.

With a view to making use of new technology opportunities, the Ministry of Education, Youth and Sport (MoEYS), with the support and assistance of the UNESCO Office in Cambodia, organized a round-table to launch a project and to formulate policies and strategies on the use of ICT in learning and education for all in Cambodia in February 2003. As a result of this national seminar, four specific policies were developed.

1. The first policy is that of ICT for all teachers and students, meaning that ICT is used as an enabler to reduce the digital gap between Cambodian schools and other schools in the world at large, especially schools in Asia and the Pacific.
2. The second policy emphasizes the role and function of ICT in education as a teaching and learning tool, as part of a subject and as a subject by itself. Apart from radio and television as a teaching and learning tool, this policy stresses the use of the computer for accessing information, communication and as a productivity tool.
3. The third policy emphasizes using ICT to increase productivity, efficiency and effectiveness of the management system. ICT will be extensively used to automate and mechanize work processes such as the processing of student and teacher records, access to information via the Internet, communication between individuals and schools, management of educational management information systems (EMIS), lesson planning, assessment and testing, financial management and the maintenance of inventories.
4. The fourth policy is to promote education for all through distance education and self-learning, especially deprived children, youth and adults who lack access to basic education, literacy and skill training, by integrating ICT with radio, television, printed materials and other media.

In line with these specific policies, the MoEYS is attempting to reduce the digital divide that exists in the different parts of the country by providing access to ICT for learning and communication to all regional and municipal/provincial teacher training institutions and then to schools across the country by 2015. The government, in its effort towards attracting investment into the field of education and human capital and to strengthen its innovation system, offers liberal incentives for investment in human capital formation, regardless of the level of investment involved. However, it appears that substantial investments are yet to flow into the field of IT training and there is the need to explore the underlying factors. Also, given the fact that there are multiple actors involved, government might consider coordinating different actors as well as begin a scheme of accreditation by NIDA or any other competent authority. It is heartening to see that some of these issues have been addressed in the draft ICT policy.

III. Trade and investment: Policies and performance

As the country emerged from political disturbances, the state was unable to commit any heavy investment and the private sector was in its infancy. A survey of industrial establishments carried out by the National Institute of Statistics in 1993 has shown that only 17 per cent of the total of 3640 establishments engaged 10 or more workers. In terms of legal organization, 62 per cent were single proprietorship, 29 per cent partnership and only 2 per cent private corporations. Government industrial establishments accounted for about 5 per cent (RGC 1996 quoted in Kannan 1997).

In the context of low rate of domestic savings (6 per cent in 1997) and weak private sector, a foreign investment law was promulgated in 1994 with a view to mobilize both capital and technology for the development of the country. This law was further amended in 2002 to make it highly liberal with a number of incentives on par with that offered by other countries in the region. The Foreign Investment law established the Council for the Development of Cambodia (CDC) as a one-stop service organization responsible for clearing the foreign investment applications. It was laid down that most of the areas are open for foreign investment with hardly any performance requirements and offered national treatment to foreign firms. However, in August 1999 a sub-decree provided for some restrictions on foreign investment in publishing, printing, radio and TV activities by limiting the foreign equity levels to 49 per cent. Also, there are certain restrictions on the use of land by the foreign nationals. Foreign investors may use land through long-term lease for up to 70 years and with possible extension.

Given the low level of availability of skilled manpower in the country, the foreign firms are permitted to bring into the country the management personnel, technical personnel and other skilled workers. The list of sectors to which investment incentives apply, without regard to the amount of investment capital, includes crop production; livestock production; fisheries; manufacture of transportation equipment; highway and street construction; exploitation of minerals, ore, coal, oil, and natural gas; production of consumption goods; hotel construction (three stars or higher); medical and education facilities meeting international standards; vocational training centers; physical infrastructure to support the tourism and cultural sectors; and production and exploitation activities to protect the environment.

Investment incentives are available for manufacturing projects in the following sectors, where investment capital exceeds $0.5 million: rubber and miscellaneous plastics; leather and other products; electrical and electronic equipment; and manufacturing and processing of food and related products. A minimum investment of $1 million applies when seeking incentives in the following sectors: apparel and other textiles; furniture and fixtures; chemicals and allied products; textile mills; paper and allied products; fabricated metal products; and production of machinery and industrial equipment.

Once the investor's application is complete and the application fee paid, the CDC is required by executive order to issue a decision on an investor's application within 28 days, although this time limit has often been exceeded.[8] Once the CDC approves the project in principle, the investor must pay a second application fee – deposit a performance guarantee – of between 1.5 and 2 per cent of the total investment capital at the National Bank of Cambodia, and register the corporate entity at the Ministry of Commerce. Once these steps have been taken, the investor will receive a formal investment license from the CDC requiring the investment to proceed within six months. Once the project is 30 per cent completed, the investor is eligible for a refund of the performance guarantee.

In general, Cambodia's investment law has been rated as one of the most liberal in the region even by the US Department of Commerce (2002d). Yet discussions with the private sector suggest that there is scope for further improvement and there exist a number of "hidden costs". To begin with, the minimum investment requirement for availing incentives might erect entry barriers for the small investors. Secondly, the need to deposit a performance guarantee and the refund of the same once 30 per cent of the project is completed might lead to a greater human interface in terms of deciding whether 30 per cent of the project is complete or not and thus providing greater scope for non-transparent practices. It is also surprising to note that in a country plagued by severe regional imbalance in development, there is hardly any provision in the investment law to attract investment into the less privileged regions. The experience of other countries tends to suggest that special provisions are needed to attract investments into the backward regions such that foreign investment acts as an agent of mitigating regional imbalance. Hence it appears that there is the need for addressing these issues in the investment law of the country.

Investment performance

Let us begin with an examination of the trend in foreign and domestic investments in Cambodia. It is evident from Table 4.2 that there has been a declining trend in both foreign and domestic investments. Notwithstanding the highly conducive investment policies, the declining trend in investment in general and FDI in particular needs to be viewed with concern. It must also be noted that the decline in FDI after the Asian crisis has been a phenomenon faced by most of the countries taken up for study. It might be possible that a large part of the investments have been made with a view to take advantage of the Most Favored Nation (MFN) status and the General System of Preferences (GSP) scheme that Cambodia enjoys with major economic powers like the US. The phasing out of the Multi Fibre Agreement (MFA) by 2005, however, is likely to diminish the present advantage for countries like Cambodia. That in turn might act as a dampener to investment in garments and textiles. There are also certain

Table 4.2 Trend in investment approvals and actual inflow in Cambodia

Year	Investment approved ($ Million)			Actual
	Total	Domestic	FDI	
1995	2,242.9	332	1,910.9	162
1996	760.8	144	616.8	586
1997	744.1	166	578.1	−15
1998	850.3	296	554.3	230
1999	447.92	260	187.92	214
2000	269.22	58	211.22	179
2001	197.71	65	132.71	113
2002	235.62	93	142.62	NA

Sources: National Institute of Statistics (2001) and UNCTAD, World Investment Report, Different Years, Geneva.

other limits to attracting investment into Cambodia because of the underdeveloped infrastructure and other complimentary inputs needed by the foreign firms.

In terms of sector-wise distribution of FDI, it was observed that the productive sectors of the economy (agriculture and the industrial sector put together) account for only about 42 per cent of the total investment (Table 4.3). Given the fact that there is enormous scope for increased investments in agriculture by bringing new areas under cultivation and the present levels of yield in agriculture are low, there appears to be the need for attracting more investment into the agricultural sector, which may be instrumental in generating employment opportunities for the rural population. But, as per the available evidence, the share of

Table 4.3 Sector-wise distribution of FDI in Cambodia (from 1994 to 2002)

Sector	Fixed assets	Share (%)
Agriculture	*356.67*	*5.73*
Agriculture	70.73	1.14
Agro industry	98.33	1.58
Cattle	2.92	0.05
Plantation	184.68	2.97
Industries	*2,317.41*	*37.26*
Anima meal	0.84	0.01
Building materials	38.79	0.62
Cement	408.49	6.57
Chemicals	7.85	0.13
Disc	2.87	0.05
Electronics	12.76	0.21
Energy	192.81	3.10
Food processing	108.75	1.75
Garment	452.28	7.27
Hat	0.88	0.01
Household goods	5.88	0.09
Leather processing	1.11	0.02
Mechanic assembly	10.28	0.17
Mechanics	0.96	0.02
Medical chemical	8.74	0.14
Medical instrument	0.08	0.00
Medical supplies	1.92	0.03
Metal	8.07	0.13
Mining	20.09	0.32
Other industries	255.71	4.11
Paper	32.67	0.53
Petroleum	85.63	1.38

Petroleum distribution	1.27	0.02
Plastics	15.97	0.26
Shoes	43.13	0.69
Socks	0.01	0.00
Textiles	84.54	1.36
Tobacco	74.76	1.20
Wood processing	441.00	7.09
Services	*1,364.93*	*21.94*
Construction	644.60	10.36
Education	100.19	1.61
Engineering	0.90	0.01
Health services	1.44	0.02
Infrastructure	188.87	3.04
Media	5.77	0.09
Service energy	0.50	0.01
Services	212.80	3.42
Telecommunication	178.85	2.88
Transportation	29.56	0.48
Water supplies	1.03	0.02
Tourism	*2,181.32*	*35.07*
Hotel	624.09	10.03
Tourism	23.46	0.38
Tourism center	1,533.77	24.66
Total	*6,220.33*	*100.00*

Sources: National Institute of Statistics (2001) and Thoraxy (2003).

agricultural sector is found to be only of the order of about 5 per cent. Given the fact that more than 80 per cent of the population still live in rural areas and depend on agriculture for their living, it is important to attract more investment into the agricultural sector.

It is found that the industrial sector accounts for about 37 per cent of total investment. Within industrial sector, highest share is accounted for by the garment industry, which has recorded the highest growth in output and employment during the recent past. It was observed that the number of garment units increased from 4 in 1994 to over 359 units in 2002. Another industry that attracts investment is leather products. The investment in these sectors seem to be with a view to take advantage of the MFN and GSP that the US, European community and other developed countries have conferred to Cambodia (Thoraxy 2003).

Trade policy and performance

Given the link between trade and investment which, as we have argued in Chapter 1, is apparently much stronger in ICT production as compared to most other industries, let us examine the trade policy framework in the country to explore the possibilities of further policy reform options, if any, to promote the production and use of ICT in Cambodia.

Changing trade policy regime[9]

The last decade witnessed major changes in the trade policies resulting in the emergence of a modern trade regime in the country. During the 1960s, Cambodia was a major exporter of agricultural products like rice, rubber and corn and the balance of payments was relatively stable. Unfortunately, the political disturbances that followed had its adverse impact on the trading system and led to the virtual collapse of foreign trade in the country. Under the trading system adopted in the early 1980s, the level and composition of trade was effectively controlled through quantitative restrictions and state-owned trading bodies, and tariffs and trade taxes played little or no role other than as a means to collect revenue. This has changed with the move to a market economy and the wide range of reforms that have since been implemented.

A process of market-oriented liberalization began in the late 1980s. The state monopoly for foreign trade was abolished in 1987 and the foreign investment law was promulgated in 1989, enabling private companies to engage in foreign trade. In 1993, trade policies were greatly liberalized. Restrictions limiting the ability of firms and individuals to engage in international trade were largely removed. There are few binding quantitative restrictions and the rates of taxes on imports and exports are for the most part not prohibitive.

The Cambodian Government continues to reform its tariff rate system. In April 2001 the number of tariff bands was reduced from 12 to 4 with the maximum tariff rate falling from 120 to 35 per cent (Table 4.4). Tariff rate reductions covered several major finished goods as well as some intermediate goods and raw materials. A study by Center for International Economics (CIE) (2001), however, argues that these changes have had a minor impact on the overall tariff structure since only a small percentage (3.8 per cent) of tariff lines had rates above 35 per cent in 2000. For example, the percentage of tariff lines duty free or subject to the minimum 7 per cent tariff rate increased from 44.3 per cent in 2000 to 44.8 per cent only in 2001 and the percentage of tariff lines with tariff rates 15 per cent or less increased from 71.6 to 73.2 per cent only. Consequently, the average tariff rate fell only slightly from 17.3 to 16.5 per cent indicating that tariffs, on average, still remained high (CIE 2001). But if we look at the issue from 1996 onwards, it is evident that there has been major reduction in tariff rates. The point may be further

Table 4.4 Cambodia's tariff rate structure

Tariff band	1997		2000		2001	
	Number	Share (%)	Number	Share (%)	Number	Share (%)
0	107	2.1	290	4.3	297	4.4
0.3	7	0.1	9	0.1		
7	2,112	40.7	2,731	40	2,758	40.4
10	14	0.3	14	0.2		
15	1,184	22.8	1,861	27.3	1,936	28.4
20	46	0.9	68	1.0		
30			4	0.1		
35	1,575	30.4	1,569	23	1,832	26.9
40			8	0.1		
50	133	2.6	256	3.8		
90			6	0.1		
120			6	0.1		
Total	5,186	100	6,823	100	6,823	100
Average tariff						
Unweighted		18.4		17.3		16.5
				(13.6)		(11.9)
Import weighted		15.9		15.4		14.2a
Effective tariff rate		NA		10.8		NA
				(12.4)		

Note: Figures in parenthesis are standard deviations, which measure the dispersion of tariff rates. Effective tariff rate is the ratio of revenue from tariffs to the value of imports. Import weighted average tariff rate for 2001 calculated using year 2000 import data.
Sources: Customs Department and Ministry of Economy and Finance, as quoted in CIE (2001).

illustrated. During 1996–2001, the imports have recorded a growth rate of over 8 per cent. Yet the tariff revenue increased only at a rate of 2.5 per cent per annum, pointing towards a significant reduction in the tariff rate.

The Government of Cambodia also made significant improvements in processes and procedures for trade facilitation over recent years. Such improvements include the following:

- It established seven public/private sector consultative working groups[10] and held four public forums chaired by the prime minister over the last two years to discuss issues raised at the working groups.
- It removed most import and export licensing requirements.

- It removed the monopoly of Caminco and introduced new legislation facilitating the entry of foreign insurers.
- It entered into a new two-year agreement in October 2000 with SGS to conduct Pre-Shipment Inspections on goods imported into Cambodia.
- It required agencies operating at border checkpoints to coordinate their activities and subject traders to only one inspection.
- It attempted to streamline procedures for issuing Certificates of Origin to garment exporters.
- It established visa-issuing facilities to individuals entering Cambodia at the major land-border crossings.

Despite these achievements significant impediments to trade remain. These impediments reflect three themes that run throughout much of Cambodian administration. These themes are

1. procedural interventions by competing government agencies, underpinned by a general acceptance of activities which supplement very low civil service salaries;
2. a lack of transparency and equitable enforcement of the law and a lack of redress from public decision making; and
3. a lack of capacity in the administration of customs and sporadic enforcement of customs law.

Absence of a healthy banking system and capital market coupled with lack of any export credit also acts as deterrence to trade and investment activity in the country. While discussing with businessmen it was discerned that bank loans are rarely available for investment and in the case of short-term loans the rate of interest is as high as 20 per cent. The underdevelopment of the banking system cannot be de-linked from the low savings rate and it could also be hypothesized that the *de facto* "dollarization" of the economy, which in turn makes the operation of the National Bank and the monetary policy in general far less effective, also has certain adverse effects. A definite conclusion, however, is not warranted in absence of a detailed enquiry.

To what extent the trade and investment liberalization has been successful in attracting investment and developing an ICT production base and promoting ICT use? It is to this issue that we turn now.

IV. ICT production and use

IT production: The present scene

Conceptually, IT production could be divided broadly into ICT goods and IT services. Each of these broad product groups comprises a wide range of goods and services with varying levels of entry barriers and incorporates varying levels of technology. Hence one of the basic premises of the present study has been that given a conducive trade and investment policy environment, even countries like Cambodia could profitably enter into some low skill and relatively stable technology areas in IT goods/IT services or IT-enabled services like data entry, medical transcription, BPO and so on at least in the medium term.

It has been inferred that there is only one firm engaged in the production of IT goods in the country. The company began its operations in 1992 as a joint venture and became 100 per cent foreign-owned by 2000. In addition to producing television sets and VCRs, it also had the dealership for leading computer companies. In the initial years the company used to employ more than 70 people. Over the years, various reasons, like high import duty and VAT (import duty plus VAT put together about 26.5 per cent) leading to large-scale smuggling and poor infrastructure, have led to a situation wherein the firm was forced to scale down its operations in the country. By 2003, the company reduced its employment to about 20 people in its IT factory and focused more on activities related to computer and software service.

Given the fact that the present level of IT production in the country is negligible, the entire domestic demand is being met entirely through imports. This has had the effect of adversely affecting the overall trade balance of the country on the one hand and forgoing the potential opportunities for employment and income generation in the country through ICT production on the other. Thus, despite the liberal trade and investment regime, IT and electronics industry could hardly attract any investment in the recent past. The fact that only very limited investment has come to this sector tends to suggest that as of now the country hardly has any specific advantage for IT and electronics production. To begin with, the local market is too small to attract domestic market–oriented FDI. Given a weak innovation system, the country lacks the complementary inputs to attract efficiency seeking investment and investment for the third country market. Thus Cambodia tends to exhibit the case of sub-optimal outcome from trade and investment liberalization on account of a weak innovation system.

IT use: Fixed telephones

Table 4.5 presents data on the trend in total number of fixed telephone lines and its distribution across different operators. Though there are three operators, the share of MPTC increased from 52.7 per cent in 1997 to 64.4 per cent in 2002. Camintel also managed to record an increase in their share from about 14 per cent in 1997 to 20 per cent in 2002, whereas that of Camshin showed a marginal decline. Despite the fact that there is apparently no monopoly, the observed annual compound growth in the number of telephone lines during the last five years has been only about 11 per cent. Analytically, the presence of more than one operator is expected to create a more competitive environment coupled with much higher growth rate. This seems to have not taken place in Cambodia. The relatively low growth rate of fixed telephone lines, however, needs to be seen against the fact that there is hardly any competition because the formation of joint ventures with MPTC has been guided mainly by mobilizing resources. Also, for MPTC being the regulator and a major participant, there is hardly any incentive to be competitive.

Not only that the overall rate of growth in telephone lines has been low, but there has also been very high regional concentration. There are three provinces (Kampot, Krong Pailin and Otdar Mean Chey) without any access to fixed telephone in 2002. Table 4.6 presents data on the distribution of landlines across different provinces in the country in 1997 and 2002. It may be noted that in 1997 both MPTC and Camshin had operations only in Phnom Penh, whereas Camintel has been providing fixed telephones in 17 provinces. Naturally, the distribution of telephone lines was highly skewed with the capital city accounting for more than 80 per cent of the total fixed lines in the country while it accounts for only 8.4 per cent of the total population. Later years saw

Table 4.5   Trend in fixed telephone lines in Cambodia

Year	MPTC	Share (%)	Camintel	Share (%)	Camshin	Share (%)	Total
1997	10,463	52.73	2,817	14.19	6,564	33.08	19,844
1998	13,835	56.64	3,436	14.06	7,157	29.30	24,428
1999	16,467	58.04	4,625	16.30	7,281	25.66	28,373
2000	18,861	62.63	5,238	17.39	6,018	19.98	30,117
2001	21,013	65.77	5,595	17.52	5,340	16.71	31,948
2002	22,206	64.47	6,898	20.03	5,340	15.50	34,444
ACGR	16.24		19.62		−4.04		11.65

Source: Estimates based on data obtained from MPTC Files.

Table 4.6 Distribution of fixed telephone lines across different provinces in Cambodia

Province	MPTC		Camintel		Camshin		Total	
	1997	2002	1997	2002	1997	2002	1997	2002
Bataeaymeanchey		115	59	208				323
Poipet		5		214				219
Battambang		310	202	638		94		1,042
Kampong Cham			145	439		176		615
Kampong Chhanag		41	48	215		25		281
Kampong Speu			38	142				142
Kampong Thom				164				164
Kampot			94	164				164
Kandal			58	218				218
Kratie								0
Krong Kaeb			103	205				205
Krong Koh Kong								0
Sre Ambel			100	123				123
Krong Pailin				49				49
Mondolkiri								0
Otdar Mean Chey				2				2
Phnom Penh	10,463	21,216	1,173	2,112	6,564	4,463	18,200	27,791
Prek Vihear				2				2
Prev Veng			37	72				72
Neak Leong		98				29		127
Pusat		43	92	261				304
Ratnakiri			67	163				163
Siem Reap		336	131	703		245		1,284
Sihanouk Ville			174	407		169		576
Stung Treng			90	149				149
Svay Rieng		21	89	242		29		292
Bavat		21						21
Takeo			117	176		110		286
Total	10,463	22,206	2,817	6,898		5,340		34,444

Source: MPTC Files.

the other two operators moving into the other provinces. Nonetheless, there appears to be no appreciable decline in the regional concentration. Viewed in terms of the fixed lines, the telecom density (lines per 1000 people) of the country as a whole increased from 0.08 in 1995 to 0.28 in 2002, where as it is almost ten times higher for Phnom Penh.

IT use: Mobile phones

As we have already seen, Cambodia has been fairly successful in attracting substantial investment, especially from Thailand, into its mobile sector and it seems to have paid rich dividend. The recorded growth rate in the mobile telephones during 1995–2002 was as high as 55 per cent as compared to a little over 22 per cent in the case of fixed lines (Table 4.7). On account of the extremely high growth rates in the mobile sector, the total telecom lines also recorded a high growth rate of over 47 per cent. The very high growth rate in the mobile sector also played a significant role in raising the telecom density in the country. While the telecom density in the case of fixed lines is as low as 0.28, that of mobile is found to be 2.54 in 2002. Hence, today there are almost ten mobile phones for every fixed line. It is also instructive to note that in terms of recorded annual growth rates, while the fixed lines show a steady declining trend, that of mobile maintain a very high growth rates, albeit there are fluctuations (Figure 4.1).

While the removal of entry barriers seems to have had the desired effect of enhancing the access to telecom, the real issue is how competitive the sector is? As we have already noted, the mobile market is dominated by Mobitel, which has recorded an annual compound growth rate of over 100 per cent during 1997–2002 and today accounts for over 53 per cent of the market share (Table 4.8). The next leading player Camshin, by recording an ACGR of over 42 per cent, improved its market share from 3.6 per cent in 1998 to over 23 per cent in the terminal year. CasaCom, the market leader of the earlier period, also recorded a relatively high growth rate of over 33 per cent, yet its market share declined from 56 per cent in 1996 to 23 per cent in 2002. Recording a

Table 4.7 Number of telephone subscribers and teledensity in Cambodia

Year	Mobile subscriber	Fixed subscriber	Total	Fixed to mobile ratio	Teledensity	
					Mobile	Fixed
1995	15,000	8,528	23,528	1.76	0.14	0.08
1996	23,098	15,475	38,573	1.49	0.21	0.14
1997	33,556	20,054	53,610	1.67	0.3	0.18
1998	61,345	24,261	85,606	2.53	0.54	0.21
1999	89,117	27,702	116,819	3.22	0.76	0.24
2000	130,547	30,877	161,424	4.23	1.09	0.26
2001	223,458	33,322	256,780	6.71	1.82	0.27
2002	321,621	35,419	357,040	9.08	2.54	0.28
ACGR	54.95	22.56	47.48			

Source: MPT Files.

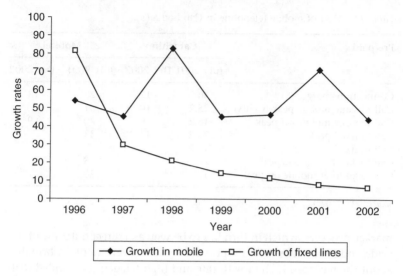

Figure 4.1 Growth of fixed vs mobile telephones in Cambodia

Table 4.8 Distribution of mobile market share in Cambodia

Year	CamShin	CamGSM	CasaCOM	Tricelxam	Camtel	Total subscribers
1996	0.00	0.00	56.30	23.18	20.52	23,098
1997	0.00	15.92	52.29	15.65	16.15	33,556
1998	3.60	45.12	39.98	5.84	5.46	61,345
1999	6.67	57.73	31.17	0.00	4.43	89,117
2000	8.08	69.66	20.40	0.00	1.86	130,547
2001	18.86	58.23	22.57	0.00	0.34	223,458
2002	23.12	53.46	22.95	0.00	0.47	321,621

Source: Same as Table 4.7.

negative ACGR, Camtel's share declined from over 20 per cent in 1996 to less than 1 per cent at the time of writing.

It appears that there is intense competition between the three leading players, which in turn led to competitive strategies like the provision of pre-paid card for a denomination as low as $5. Also there has been significant reduction in the price of mobile telephone. For example, the cost of a local call from mobile to mobile as well as the connection charges have come down by more than 50 per cent during the last two years (Table 4.9). On the whole, the role of highly competitive

Table 4.9   Cost of mobile telephone in Cambodia ($)

Pre-paid	Camshin		Mobitel	
	July 2001	Dec 2002	July 2001	Dec 2002
Connection charges ($)	30	12	22	13
Call to same mobile peak (Cents)	25.2	10	31	12
Call to same mobile off peak	15.2	9	20	9
Call to fixed peak	25.2	11	33	
Call to fixed off peak	15.2		20	
Call to another mobile peak	28		33	
Call to another mobile off peak	16		20	

Source: ITU (2002c).

market structure, which in turn has to be seen as a direct outcome of the trade and investment policies pursued by the government, in bringing about the recorded high growth rate and high teledensity coupled with declining prices in a poor country with very low per capita income ($260) cannot be over emphasized. But it should be noted that given the present state of technology there are obvious limits set by the mobile technology to enable the people to utilize the power of Internet. Hence the important issue is whether the lower level as well as the declining rate of growth of fixed lines has any implications on harnessing ICT for development. This takes us to an examination of the state of Internet in the country.

Internet and computers

By June 2001 there were only about 8000 Internet connections in the country. Of these connections Camnet had about 1796 subscribers (with only one leased-line subscriber), which were distributed as follows: private customers 65 per cent; commercial enterprises, 28 per cent; educational institutions, 3 per cent; and government, 4 per cent. In the case of Bigpond, out of its 2008 subscribers 90 per cent were in Phnom Penh and 10 per cent in Siem Reap. It had seven leased-line subscribers including WHO, some NGOs and MobiTel.

Notwithstanding the drastic reduction over the years, the Internet tariff at the time of writing was high in relation to the income levels in the country. When Internet was first introduced in the country, Camnet charged as much as $8 per hour. As is evident from Table 4.10, the tariff levels at the time of writing were only a fraction of what prevailed in 1997. Yet as ITU (2002c) rightly pointed out, Cambodia has the dubious

Table 4.10 Internet pricing in Cambodia: an illustration from Camnet

	Deposit		Monthly fee		Free hours		Extra hour	
	2001	2003	2001	2003	2001	2003	2001	2003
Option 1	40	30	18	15	3	6	3.4	2
Option 2	40	30	30	30	8	13	2.7	1.8
Option 3	40	30	90	90	30	45	2.45	1.4
Option 4	40	30	300	180		100	2.25	1.2
Option 5	40	30	600	NA		3	2.10	2.4

Note: Data for both years refers to the month of September.
Source: ITU (2002c).

distinction of having the lowest Internet penetration in Southeast Asia as well as the highest prices.

Cambodia has very few leased-line subscribers, mainly on account of the exorbitant price. Camnet charges a flat rate of $3500 per month for a 64 kbps leased line whereas the cost of Bigpond's leased line depends on the utilization of the line. It charges a monthly rent of $250 and in addition a usage fee in the range of $1850–$4150 depending on the usage. One Mbps line from Camnet would cost as much as $48,500 per month!

The UNESCAP (2002) finds that access to IT infrastructure in the government sector is extremely low. There are about 15,000 people in the government with access to computers. Most of these computers, though networked, do not have access to Internet leading to a situation wherein Internet access is limited to only about 1–2 per cent of the government staff. Though there are over 20 ministries, only a few have web presence.

V. Concluding observations

To sum up the discussion, the present state of IT use and production tends to suggest that the trajectory has been in tune with the overall development of economy. While there is some merit in the argument that the country is caught up in the choice between Pentium and Penicillin, one has to take into account that in Cambodia the number of passenger cars is almost four times that of neighboring Vietnam. Even though the affordability is an important issue to be taken up for discussion below, the demand side argument might not provide an appropriate answer. This takes us to the supply side of the issue, which is inexorably linked to the technology and innovation system, which was in its primitive stage at the time of writing. The technological choice that the country has made with respect to telecommunication has led to a situation wherein the number of fixed lines in the country has remained stagnant over the years. Thus, there appears to be some merit in the argument that limits to Internet expansion in the country have been set by the availability of fixed lines. This in turn may be seen as an outcome of two factors: first, the fixed-line market has been to some extent a quasi-monopoly of MPTC with very limited resources to invest for its expansion. Secondly, the government, as early as in 1993, opted for mobile technology instead of fixed lines to meet the telecommunication needs of the people and also opened it to the private sector. Now, there appears to be two options worth considering: first, to do away with the state monopoly in fixed lines and to explore the possibilities of attracting more investment into this area. Secondly, to induce the mobile firms to go up the technological trajectory and make possible the data transmission through mobile phones. Equally important is the weak institutional arrangements for the development and diffusion of ICT and human capital base that in turn set limit not only for the use but also for the production of ICT. It is found that today there are a number of actors involved in dealing with different IT-related issues and there is the need for a separate ministry to coordinate the activities of different agencies involved. In any case, the ICT policy makers in the country will have to deal with these issues in the near future.

The above line of argument does not imply that the issues of access and affordability are not important in the case of Cambodia. The issue of affordability arises mainly on account of the high price of hardware and software in relation to average income level of people in countries like Cambodia. The developments in Free/Open Source Software (FOSS), though in its early stages of development, are likely to provide an alternative for countries like Cambodia. This, however, is not an issue specific to Cambodia. Hence there appears to be enormous scope for joining hands with other less developed countries to find ways and means to address the issue of affordability.

Given the fact that the production base for IT goods is practically absent in the country and this could be instrumental in providing new income-earning and employment-generating opportunities, concerted efforts are called for to develop an ICT production base in the country. In the sphere of ICT services there are opportunities to enter into some of the relatively less skill-intensive ICT services like data entry and IT-enabled services like medical transcription, call centers and so on.

It is also to be noted that the skilled labor for such services could be developed in the short run and they are ideal for generating large-scale employment. But such IT-enabled services also call for better communication infrastructure at affordable prices. Hence the present study underscores the need for initiating steps such that Cambodia is able to find a place in the international production networks of IT in the near future.

The government has been fairly successful in creating an environment conducive for the growth of private investment in the country. Incentives offered and the procedural reforms are on par with other countries in the region. Yet the study finds that there are certain issues that might stand in the way of attracting investment into the country and in promoting a balanced regional development. First, the policy of specifying certain minimum investment requirement for availing incentives might erect entry barriers for the small investors. Secondly, the need to deposit a performance guarantee and the refund of the same once 30 per cent of the project is completed might lead to a greater human interface and thus leading greater scope for corrupt practices. It is also surprising to note that in a country plagued by severe regional imbalance in development, there is hardly any provision in the investment law to attract investment into the less privileged regions.

In making efforts towards developing an IT production base, it is important to keep in mind the lessons offered by the experience of other countries. To begin with, the strategy might be to make available a large pool of IT manpower at different levels such that the primary condition for the establishment of IT goods/service production base is satisfied. Here the strategy needs to be one of pooling together the resources of different actors like Civil Society Organizations, private sector and so on. Also the strategy should not be one of spreading thinly the resources across the country, instead the investment needs to be undertaken in such a way as to take advantage of the agglomeration economies. This might be possible through the setting up of technology parks wherein built-up space, communication infrastructure and others, which are beyond the reach of an individual entrepreneur, are provided along with a "single window clearance" system so that the prospective investors need to have only limited interaction with the bureaucracy. Secondly, such technology parks need to be close to and have constant interaction with the centers of learning such that mutual learning and domestic technological capability is built up in the long run. Thirdly, there is also the need for conscious efforts towards skill empowerment and learning process such that the economy does not get locked up in low technology activity and an upward movement along the skill spectrum is ensured. It needs to be noted that in the investment policy of Cambodia, on-the-job training has not received the attention that it deserves. On the whole, it appears that given the specific characteristics of ICT sector there appears to be the need for a balanced approach wherein liberal trade and investment regime is accompanied by concerted efforts towards building up an innovation system.

5
Lao PDR: Hastening Slowly?

I. Introduction

Notwithstanding, more than a decade of concerted efforts towards liberalizing the economy and attracting investment towards the productive sectors of the economy and resulting annual growth rate of over 5 per cent Lao PDR, even today, exhibits symptoms of an underdeveloped economy. Over 80 per cent of its population lives in rural areas depending on agriculture and allied activities. Yet, they contribute only about 50 per cent of the GDP pointing towards low levels of productivity in the primary sector. The industrial sector, which contributes about 23 per cent of the GDP, essentially, comprises primary processing and is characterized by high regional concentration. In 2001, Vientiane Municipality and Vientiane Province together accounted for 55 per cent of the large, 41 per cent of the medium and 19 per cent of the small industrial units[1] and over 41 per cent of the industrial employment in the country. The service sector, which contributed almost a quarter of the GDP in 2001, was found dominated by wholesale and retail trade. This however, does not imply that there has not been any structural change in the economy. The observed change during the last decade mainly was in terms of a decline in the share of agriculture from about 61 per cent in 1990,[2] the initial year for which data is available, to a little over 51 per cent in 2002. Correspondingly, the share of industrial sector increased from 15 to 23 per cent while that of the services sector recorded a marginal increase from 24 to 26 per cent. Within the services sector, there has been a remarkable increase in the share of hotels and restaurants from about 0.1 per cent in 1990 to 8.5 per cent in 2001.[3]

Since the launching of New Economic Mechanism (hereafter NEM) – which emphasized greater role of FDI, privatization of state-owned enterprises[4] and greater integration with the world market – there has been a turnaround in the country's output growth. During 1992–97 the recorded annual average growth rate in GDP was of the order of 7 per cent. The high GDP growth was also accompanied by corresponding decline in the incidence of poverty from about 45 per cent in 1992 to a little over 38 per cent in 1998. This was followed by a slow down in the pace of growth on account of the Asian financial crisis that badly hit the Thai economy, which is the major source of exports and investment in Lao PDR. Annual output growth almost halved, year-to-year inflation rate rose to more than 150 per cent in the first quarter of 1999, foreign investment fell almost 70 per cent in 1997–98 and the Lao currency lost about 90 per cent of its value in the 18 months during the period ending August 1999. If the recent data is any indication, the economy appears to be recovering from the crisis, *albeit* slowly. The recorded growth rates in 2001 and 2002 have been a little over 5 per cent.

Highly dependent on Thailand for both exports and imports and international donors for aid, the Government of Lao PDR recognized the need to integrate further with the regional and international economy. Lao PDR joined ASEAN and AFTA as a full-time member in July 1997, and applied for membership in the WTO. Lao PDR is still in an early phase of the accession process but looks forward to the entry into WTO in the near future. The government's efforts towards integrating with the world economy have also coincided with the series of initiatives towards harnessing ICT for development (Thampi 2003).

II. Innovation system in the ICT sector

Policy initiatives

In 1996, the Science Technology and Environment Agency (hereafter STEA) was given the approval by the prime minister's office to implement the overall policy for monitoring and controlling IT in the Lao PDR. Given this mandate, STEA proposed a four-year plan (1996–2000) which dealt with, among other things, developing an IT infrastructure (including human capital, IT industry base, communication network), promoting IT application (in government, business and industries as well as economy at large) and devising policies for promoting IT development which also included policies relating to promotion of FDI in IT. Achievements by the year 2000 appeared to have fallen short of the targets on account of the lack of resources, lack of coordination among different agencies involved, low levels of IT literacy and lack of infrastructure and so on (UNDP 2002). Later the government initiated steps towards formulating an integrated IT policy for the country, and five working groups have been appointed for this purpose.

In terms of institutional interventions, an IT Center has been set up under the Ministry of STEA (www.stea.gov.la). Apart from providing training to government officials in various departments, the Center also undertakes research on certain issues specific to Lao PDR, such as open source software,[5] building Urban–Rural Digital link, development of Lao character set and e-governance. While International Development Research Center (IDRC), Canada, provides assistance for Rural–Urban digital link and the project on Lao character set, Government of India provides assistance for e-governance.

IT infrastructure: Radio and television

Television and radio are considered as two most important means of mass communication in developing countries. In Lao PDR, television is relatively new, but it has already seven channels from five stations at the local level and two stations at the national level. The five government-owned television stations broadcast daily in Lao and the regional stations broadcast in Lao and in tribal languages. In 1990 a new satellite ground station was installed and the next year agreements with China and France were concluded to relay their broadcasts to Lao PDR by satellite. TV programs from Thailand received in the southern part of Lao PDR are subject to censorship. Given the insignificant number of TV owners and transmission difficulties caused by the country's mountainous topography, cable TV seems to be a rather expensive and unprofitable venture and it is yet to begin in the country. When it comes to diffusion of television, level of 10 sets per 1000 people (in 2002) was also very low by any standard and there were only a dozen of least developed countries with television density less than Lao PDR.

There are radio stations at Luang Prabang, Siphandon, Parkse, Houaphan, Savannakhet and Vientiane. In addition, FM stations are also located at Attopeu, Luang Namtha and Saiyabouli. The state-owned radio service has a national network and broadcast in Lao and tribal languages.[6] The number of radios per 1000 people is only 143. While it is better by least developing country standards (in 2002, 22 countries were found having radio density lower than Lao PDR), it is at a very low level by international standards.

Telecommunication: Fixed telephones

Telecommunication services, especially the fixed telecom services, may be considered as the key factor in ICT for exchange of information, especially that of large volumes of data. Hence any attempt at harnessing ICT for development needs to begin with the provision of telecom facilities. A study which analyzed the link between access to telephone and demand for health services in the three emerging market economies, including Lao PDR (others being Peru and Bangladesh) has concluded that if rural development and health improvements were to continue to be the objectives, then the development of telecom services in rural areas appears to be a factor powerfully supporting these objectives (Micevska 2003).

The realization of the paramount importance of telecommunication in the development of Lao PDR and the commitment of the government towards providing access to Internet is evident from the Article 4 of the Telecommunication Act of the Lao PDR which states that the state encourages local and foreign investors to compete and cooperate in investment in the construction, development and expansion of telecommunication network and services in accordance with the system prescribed by the government. Further, there has been a series of initiatives undertaken by the government during the last decade to provide telecom access to people. In Lao PDR, the Department of Posts and Telecommunications within the Ministry of Communication, Post and Construction (MCPTC) has been responsible for the formulation of telecommunication policies and regulating the behavior of different participants. From the early 1990s, there have been a series of initiatives like facilitating greater private sector participation to raise resources for the expansion and modernization of the telecom sector on the one hand and to bring about a competitive environment and provide access to people at affordable prices on the other as shown in Table 5.1.

By 2003 ETL of the Government of Lao PDR, LTC, Lao Telecom Asia Co. Ltd (LAT) and Milicom Lao Co. Ltd (Tanko) were providing telecommunication services throughout the country. Out of the 18 provinces, 5 provinces had been provided with fiber-optic network connectivity in their capitals. ETL had been working towards a target of providing fiber-optic network in rest of the 13 provinces by the end of 2004 and 142 district headquarters by 2005.

Computers and Internet

Given the fact that telecommunication and computers are the two integral parts of Internet, the level and rate of diffusion of Internet is

Table 5.1 Milestones in the development of telecom sector in Lao PDR

Year	Milestones/Initiatives
1990	Entry of Private sector, in the hitherto fully state-owned telecom sector, for the first time – Telestra, Australia, began telecom operations
1993	Beginning of Cellular network in Vientiane
1994	Formation of Lao-Shinawatra Telecom Company – of a Joint Venture between Shinawatra Thailand and Government of Lao PDR, christened as Lao-Shinawatra Telecom Company (LST)
1994	Construction of microwave transmission system covering the north to south of the country
1995	Separation of telecom department from the postal department that is the Enterprise of Post and Telecommunications Lao (EPTL) was divided into Enterprise of Post Lao (EPL) and Enterprise of Telecommunications Lao (ETL)
1996	Introduction of public telephones (200nos) by LaoTel to provide wider access
1996	Merging of ETL and LST to form Lao Telecommunications Company (LTC also called LaoTel), keeping the majority control (51 per cent) with the Government of Lao PDR . Also having monopoly in fixed and mobile services for five years
1999	Expansion of the public telephones to make the total number to 300
2000	With a view to avail Japanese assistance under ODA to modernize the telecom facilities, ETL was re-established
2000	Beginning of pre-paid services
2001	ETL initiative in massive expansion of Telecom infrastructure by the construction on a fibre optic backbone, coving whole of the country
2002	End of Lao Tel monopoly and the entry of ETL into Telecom services market resulting in increased competition
2002	Entry of foreign firm, Millicom, as a joint venture with the government into mobile services

Source: Compiled by the author from UNDP (2002), ITU (2003), UNESCAP (2001) and discussion with different government officials in Lao PDR.

likely to be affected by the availability of telecom infrastructure, cost of its access, access to computer and its cost. If the available evidence is any indication, the Government of Lao PDR is aware of the importance of Internet for the overall development of the country and its own role in providing Internet access to people. It is evident from the Directive on the development and use of ICT in Lao PDR (Directive No. 14 dated 21 December 2001). The objectives of the directive were

- to facilitate wide and effective use of ICT in every sector as an essential factor for the socio-economic development, the national defense and security;
- to develop a national telecommunication network for high-speed communication and increased access to the Internet country wide;
- to promote competition in the ICT network and services;
- to broaden and enhance international cooperation, to create a favorable condition for the development and use of ICT through the national telecommunication network infrastructure; and
- to develop human resource in the field of ICT as a key factor for the successful implementation of the ICT.[7]

The commitment of the state is further evident from the different initiatives made by the state during the last decade as presented in Table 5.2.

Table 5.2 State initiatives towards promoting Internet in Lao PDR

Year	Initiatives
1994	Beginning of Internet at the instance of Lao expatriates and others to form Laonet, but later this was discontinued
1996	Order of the office of the prime minister to authorize the Science Technology and Environment Agency to implement the overall policy for monitoring and controlling IT in Lao PDR
1996	Formulation of IT Master Plan (1996–2000) by STEA dealing with different aspects of IT development in the country
1998	Formation of Globenet set-up by an American expatriate to provide Internet services using a satellite, operating via Philippines
1998	Australian-owned Planet café, largest Internet café in Lao PDR, begins its operations
1999	Entry of PlanetOnline (fully foreign-owned) into the Internet market
1999	Beginning of the internet Services by LaoTel – a joint venture between Lao Government and Shinawatra
2001	Government Directive (No. 14) on the development and use of ICT in Lao PDR
2003	Introduction of one Internet gateway system for Lao PDR (dot.la) with a view to monitor and regulate all Internet activities in the country
2003	Setting up of five different task forces to draft a new IT policy for the country

Source: Compiled by the author from UNDP (2002), ITU (2003), UNESCAP (2003) and based on the discussions with government Officials.

The table clearly indicates that there have been bold initiatives towards improving the Internet access by promoting investment and creating a more competitive environment. As of May 2003, Internet access in the country was provided by seven ISPs of which six of them were in the private sector. The government-owned Lao National Internet Committee (LANIC) confines its operations to provide Internet services to the government organizations and academic institutions.

Human capital

So far we have dealt with the physical infrastructure related to ICT use. But even if a modern ICT infrastructure is developed and offered at affordable prices, the real benefits may not be accrued to the rural masses unless the content is developed in accordance with the needs of the people, and the people at large have the capability to use them. Here lies the need for human capital building for producing the required content and also for making effective use of the new technology.

As of now, computer and IT-related education[8] in Lao PDR is provided not only by the National University of Lao PDR, but also by the private sector. The computer science program in the National University of Laos (NUL) began in 1998 under the Faculty of Science, Department of Mathematics.[9] The number of outturn of graduates is only 29 (UNDP 2002). The faculty of Engineering and Architecture (FEA) is considered as best equipped with IT facilities in the NUL system.[10] The main component of these facilities is the Lao-Japan Technical Training Center (LJTTC). The courses offered at the LJTTC are a combination of general application courses: computer-aided engineering courses and a course on network software. LJTTC even offers a course on Internet café set-up and maintenance.

Given the limited IT education facilities in the public sector the vacuum is filled up, at least partly, by the private sector. The following private colleges are currently providing IT education, albeit at a very preliminary level. The Vientiane College, a private institution with the academic and financial support of the Monash University in Australia, was established in 1992. Other institutions involving foreign investment are the Micro Info Centre (Joint Venture) and Lao American College, another Joint Venture. The Lao American College has established working relations with the National University of Lao PDR, City University of Washington State, USA, the Ohio University, USA, and the Bangkok University, Thailand. In addition to these educational institutions with foreign investment, there are three local initiatives, namely Rattana Business Administration College, Com Centre and PVK Computer Center. Parallel to these educational institutions, there are a number of computer dealers who provide short-term training in computer operations. Given the multiplicity of actors involved, from the long-term interest of creating a high quality human capital stock, it may be worth considering an accreditation scheme at the instance of STEA.

There is an active cooperation between the Governments of India and Lao PDR to develop the IT sector in general and human resources in particular. Government of India assists the Government of Lao PDR in establishing an "Information Technology Center with Value Addition" at Vientiane under its International Technical Exchange Cooperation (ITEC) Program. A Memorandum of Understanding (MOU) on Information Communication Technology Cooperation with Government of Lao PDR envisages the following:

1. setting up of an ICT Training Laboratory with 25 Computers with software tools;
2. capacity building of government officials through Human Resource Development for 150 government officials in four Course Modules of 5-days duration;
3. MCA Program for 30 Lao Students in India of 3-year duration;
4. setting up of National Data Centre (e-Governance Infrastructure);
5. establishment of five pilot rural telecenters (Community Information Centre) in remote and rural areas along the Mekong River for ICT Penetration in agriculture, rural health and rural development programs;
6. establishing a VSAT-based network for 18 Provinces Governors (Provincial Informatics center – PIC);
7. assisting in preparation of Cyber Security, Cyber Laws and certification authority;
8. entrepreneur development programme in ICT quality education;
9. solving the issue of Lao Language Information Processing.

Under capacity building, training has been imparted to 150 government officials of Lao PDR. An Entrepreneurship Development Centre (EDC) with the support of Government of India has also been set up in Vientiane.

Government is also highly committed towards enhancing the IT capability of its employees. The present approach is one of "Training the Trainers" wherein selected number of officers is provided with an opportunity to undergo training in computer-related courses. These officers, in turn, are expected to impart training to their colleagues once they return to their parent office. Such a strategy presupposes that anybody who has undergone training will be "willing and able" to train others.[11] To the extent that such a presumption may not turn out to be correct, the outcome will be less desirable. Hence, it is important that facilities are created within the country, perhaps in association with those renowned universities/institutes in the neighboring countries like India.

In a society wherein the younger generation considers "Internet and English as their lifeline", it is surprising to find that as of now there are hardly any primary or secondary schools in the country which provide computer access to students or impart any kind of computer education.[12] It is has been found that the ministry of education, which has some 140 computers, of which 40 are Internet compatible, has a "top-to-bottom" approach wherein the ministry first develops ICT within the ministry itself, then within the university and only finally within the schools (ITU 2003). In this context, it may be advisable for the ministry to evaluate the social marginal product of "bottom-up" approach *vis a vis* the "top-to-bottom" approach. We would suspect that the return is likely to be much higher in the former as compared to the latter. It is important to note that adequate supply of IT manpower is not only needed for the generation of Internet content and effective use of Internet, but it offers an opportunity for the country to enter into the fast growing area of IT-enabled services. Thus, the need to promote investment in the area of IT manpower development cannot be overemphasized.

III. Trade and investment regime

Evolution of policies governing FDI and present scene

Originally, the Investment Promotion Law (IPL), which has been in force since 1989, governed the inflow of investment into the Lao PDR. In 1994, a revised Foreign Investment Law was passed with a view to provide a more propitious environment for FDI. It contained a series of liberal measures designed to attract capitalist-style enterprise like repatriation of profits and the involvement of foreign equity in Lao businesses. It outlined the areas in which investment has been encouraged and those areas where investment was not allowed. Prohibited areas included activities detrimental to the environment, public health or national culture. An important part of the revised IPL related to the types of investment. Before the revision, there existed three types of investment – joint investment, wholly foreign investment and investment under contract. The third type was removed due to its complexity, which confused investors.

In 2001, as per the Decree of the Prime Minister (NO. 46/PM) regarding the Law on the Promotion and Management of Foreign Investment in the Lao PDR, further changes were effected. The law required special presidential approval for any investments relating to natural resources, environment, public health and national culture. Although there are no other restrictions on activities permitted to foreign investors, prohibited sectors of business are not well defined, leading to a situation wherein the investors need to conduct enquiries before beginning business. The revised law reduced the approval steps required for investment projects to just one, which is known as the one-stop-service system, and the duration of the examination of the project was decreased from 90 to 60 days. Meanwhile, registration of the enterprise must be completed within 90 days – half the time required before the revision. Of late it has been further reduced to 15 days in the case of projects where the investment is less than US$1 million.

The new law also recognizes the ownership of capital, property and interest of foreign investors and offers national treatment. The law further states the forms of acceptable foreign investment and the rights, benefits and obligations that come with such investment. It also explains the responsibilities of the newly formed Department of Domestic and Foreign Investment (DDFI), the government body that deals with inward investment.

The FIL permits two forms of foreign investment: wholly foreign-owned enterprises and joint ventures. A wholly foreign-owned enterprise is a foreign investment registered under the law and regulations of the Lao PDR by one or more foreign investors, without the participation of domestic Lao investors. The enterprise established in the Lao PDR may be a new company; alternatively it may be a representative office of a foreign company. Banks have the option of establishing branch offices though they are restricted to Vientiane. The foreign investments must have a minimum registration capital of US$1,00,000. The license for a wholly foreign-owned enterprise will have a maximum life of 15 years. This can be extended if approved by the DDFI.

The stipulation of minimum capital requirement of $0.1 million is likely to have the effect of erecting entry barriers to certain foreign enterprises; especially the small- and medium-sized ones. This in turn could have an adverse effect on the total investment inflow. This is because, in the current era of globalization, it has been observed that large multinational firms are not the only source of investment resources, management expertise and technology that is badly needed by the developing countries like Lao PDR. But, there are a large number of SMEs having the financial and other resources and keen on investing in the developing countries. In the Indian context, for example, an overview of the foreign collaborations approved during the post-1991 period reveals that in each year there were a large number foreign investment proposals involving investment less than $0.1 million. For example, in the year 1999, out of the 1352 financial collaborations approved the share of cases involving foreign investment less than $0.1 million has been as high as 30 per cent. Hence, opening the doors of investment with less than $0.1 million might be instrumental in attracting more investment to Lao PDR. Such an argument holds especially for the less capital-intensive IT service sector.

In a joint venture, foreign investors must contribute at least 30 per cent of the total equity investment. The license will have a maximum life of 20 years. This can be extended if approved by the DDFI. Foreign investments involving exploitation of natural resources and energy generation must be joint ventures. Here again, setting the minimum foreign share might act as an entry barrier. Hence it may be worth considering the economic rationale for setting the minimum contribution at 30 per cent, for such a minimum contribution appears to be not a necessary condition for reaping the benefits of foreign investment. Perhaps it may be better to leave to the foreign and local counterparts to determine the sharing of investment. Such an approach is also likely to enhance the invest inflows to Lao PDR. Here it may be worth noting that in 1999 almost 9 per cent of the financial collaborations approved in India were involving less than 25 per cent foreign equity.

Lao PDR does not impose performance requirements like minimum local content *per se*. At the same time, for obvious reasons, foreign investors are encouraged to give priority to Lao citizens in recruiting and hiring. Foreign firms are also permitted to hire foreign personnel if necessary. But, before bringing in foreign labor, the enterprise must apply for work permits from the Ministry of Labor and Social Welfare. A list of foreign personnel must also be submitted to the Investment Service Center of Foreign Investment Management Committee (FIMC).

The FIL provides a number of incentives which are comparable to other neighboring countries (Freeman 2002). In Lao PDR a single rate of 20 per cent annual profit tax is applicable for foreign investments, which is much lower than that in Vietnam or in Thailand. Import duty is imposed on imports of equipment, means of production, spare parts and other materials used in the operation of foreign investors' projects or in their productive enterprises at a uniform flat rate of 1 per cent of the imported value. Raw materials and intermediate components imported for the purpose production for exports are exempt from such import duties. All exported finished products are also exempted from export duties.

In highly exceptional cases, and by specific decision of the Lao Government, foreign investors may be granted special privileges and benefits. These can include a reduction in or exemption from the 20 per cent profit-tax rate and/or 1 per cent import-duty rate. Such reductions and exemptions are normally given because of the large size of an investment and the significant positive impact that it is expected to have upon the socio-economic development of the Lao PDR. On the whole, going by the declared policy on FDI, it may be inferred that there have been significant changes in the policies governing inflow of investment. At the same time, there appears to be room for improvement as is evident from the following statement.

While Lao PDR' body of commercial law is slowly developing, foreign investors most frequently cite inconsistencies in the interpretation and application of existing laws as among the greatest impediments to investment. The lack of transparency in an increasingly centralized decision-making process, as well as the difficulty encountered in obtaining general information, augment the perception of the regulatory framework as arbitrary and inscrutable. Moreover, there is a feeling among the investors that the red-tape requirements associated with establishing a foreign investment have actually proliferated. (US Department of Commerce 2002b)

Trend in FDI: Approved and actual inflow

To the extent that there have been significant changes in the policy environment, the impact of the same may be gauged by examining the extent to which these reforms have been instrumental in attracting FDI. But while analyzing this one cannot forget the fact that the today's least developing countries are faced with a significantly more difficult international environment as compared to their counterparts in the 1970s and 1980s. In the 1970s and early 1980s the developing countries, in general, were not friendly, if not hostile, to the foreign investment. In such an environment, the countries which had opened their doors for foreign investment were fairly successful in attracting substantial investment. The late 1980s and 1990s engendered a major change of heart among the policy makers of developing world – with the policy

Table 5.3 Trends in the approved and actual inflow of FDI to Lao PDR (US$ million)

Year	No. of projects	FDI approved	Actual FDI inflow	Domestic investment
1990	43	90.38	6	247.25
1995	63	534.24	88	270.67
1996	63	972.18	128	320.47
1997	62	113.61	86	25.17
1998	69	97.78	45	24.75
1999	60	108.13	52	31.01
2000	68	190.33	34	58.76
2001	63	1017.05	24	326.09
2002	43	277.7		

Sources: National Statistical Center (2000, 2002) and UNCTAD (2004).

pendulum swinging from import substitution to outward orientation. The result has been, as UNCTAD (1995) noted, intense competition among developing countries to attract FDI. This has resulted in what is called the "incentive competition" between developing countries, which in its ultimate analysis appears detrimental to these countries.

Table 5.3 presents data on number of projects approved involving FDI and the local investment involved in these projects. We also present data on actual inflow of FDI. The table indicates that there has been a marked decline in the FDI approvals after 1996 although in terms of the number of projects approved there has not been any marked decline. Nonetheless, viewed in terms of approvals, there appears to be a turn-around since 2000. It is also evident that the rate of FDI fructification (defined as the ratio of approved FDI to actual inflow) has been significantly low. For the period under consideration, the fructification rate was found to be as low as 14.8 per cent. Now given the fact that there has not been any marked decline in the number of projects approved, whereas the fructification rate has been rather low, it may be inferred that the MNCs find it difficult to obtain the complementary inputs notwithstanding the liberal policy regime and incentives on par with other countries in the region. Hence, there appears to be no other option but to strengthen the innovation system such that Lao PDR emerges as an attractive location for FDI.

Link between foreign and domestic investment

Apart from the issue of FDI fructification, which calls for simultaneous focus on investment promotion and implementation, there appears to be yet another issue that emerges from the table. This pertains to the relation between foreign and domestic investment. In an economy, generally perceived as having a number of local impediments to overcome for investment, it is only natural that the foreign investors would like to join hands with the domestic capital. This is because of the comparative advantage of local capital in handling issues relating to obtaining required licenses and overcoming specific hurdles in the process of project implementation. As one of the successful entrepreneurs in Lao PDR, who is a partner in a joint venture, remarked, "in my absence my foreign counterpart would not have ventured to invest in Lao PDR". Hence there is some basis to believe that domestic capital has a so-called "crowding-in effect" on foreign investment.

With a view to empirically explore, in a rather rudimentary manner, the perceived relationship between the foreign and domestic investment we have plotted the data on domestic and foreign investment on a graph (Figure 5.1) It is evident that there is a synchronization in the movement of foreign and domestic investment. Hence it appears that the presence of a strong and thriving domestic private sector acts as a signaling device to the foreign capital.[13] It has also been found that mergers and acquisitions dominate FDI inflows into developing and transitional economies (UNCTAD 2000). Therefore, any attempt towards promoting FDI, to have the desired outcome, has to be complemented by simultaneously promoting the private sector. In an economy

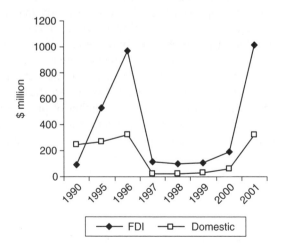

Figure 5.1   Trend in approved foreign and domestic investment in Lao PDR

like Lao PDR, which has been earlier under the socialist regime, it is only natural to expect the domestic capital to be in its infant stage. It is therefore not surprising to note that the share of private sector in GDP is only around 10 per cent and most of the large enterprises are in the capital city with small and medium firms operating in the provinces and rural areas. This being the case, a careful examination of incentives offered by the government indicates that there appears to be a discrimination against the domestic capital because they get much lower concessions than their foreign counterparts. To illustrate, the income tax rate for the domestic capital (30 per cent) is much higher than the rate for the foreign capital (20 per cent).[14] Hence it is important that a level playing field is offered for both domestic and foreign capital.

Given the fact that the entrepreneurs are "not only born but also developed", special emphasis may be given, preferably at the instance of the Department of Promotion and Management of Domestic and Foreign Investment, to create and nurture an entrepreneurial class in the country. The Entrepreneurial Development Programs (EDPs), as have been carried out in India mainly at the instance of Entrepreneurship Development Institute of India, Ahmedabad (in the state of Gujarat),[15] could be a possible model to follow with due adaptation to local conditions. Thus, in addition to the efforts to generate skilled manpower in general (which we have discussed in the previous section) there appears to be, for historical reasons, the need for focused efforts towards entrepreneurship development, which is bound to provide rich dividends in the years to come. In this process again the present study underscores the role of multilateral organizations.[16]

Investment in IT

Our enquiry has shown that as of now there is hardly any investment, either foreign or local, in Lao PDR for IT production. Yet, the government has been successful in attracting some investment (both foreign and domestic) into the telecom sector as well as for IT manpower generation. As already stated, the investment policy encourages private investment in telecommunication and in the field of IT training. Though there is no explicit policy towards attracting, investment in the field of ICT, according to the decree of the prime minister (No. 46/PM), the following areas in computer and related activities like

- hardware consultancy
- software consultancy and supply

- data processing
- maintenance and repair of office, accounting and computing machinery

and areas relating to human capital formation like

- general education
- technical and vocational secondary education and
- adult and other education

are open for foreign (as well as domestic) investment without any restrictions. This clearly indicates that given the strong linkage effect that IT and human capital on the overall development of the country, these areas are considered as priority areas for investment. However, substantial investments are yet to flow into these sectors. While the limits set by small domestic market is important, it appears that the weak innovation system as manifested in lack of human capital, vibrant local entrepreneurial class along with poor financial sector, would have played its role.

Trade policy framework and trade facilitation

In 1986 the Lao Government launched the NEM, with the ultimate aim of turning the Lao PDR into an open-market economy. Initial progress was rapid, as manifested in the series of reforms to open up the economy for FDI, and removing many of the man-made restrictions on trade. This greater integration with the world economy facilitated the entry of Lao PDR in the ASEAN in July 1997. It is in the process of bringing about further lowering of its tariff barriers in line with the goal of complying with the tariff reduction schedule of the AFTA by 2008. Lao PDR also has an observer status within the WTO and applied for full membership in 1997. The ongoing trade and investment liberalization has to be seen against the fact that WTO membership is conditional to its compliance with WTO trade regime.

As of now, Lao PDR has a simplified tariff structure, although some non-tariff barriers, such as a quota on the import of automobiles, still exist. Given the widening trade gap, additional restrictions on the import of certain luxury goods are currently under consideration by trade authorities. Importing from and exporting to Lao PDR still requires authorization from several national and local authorities, which can be a time-consuming and less-than-transparent process. At the same time, it may be noted that the computerization of customs department with UNDP assistance has seen major improvement in terms of a marked reduction in the time taken to clear imports from about two days to about an hour.

As in many other developing countries, Lao import tax system has been framed in such a way as to promote the import of materials and equipment for investment and production, while protecting domestic production and limiting luxury imports to conserve limited foreign exchange. Also, in an economy with growing fiscal deficit, import tariff is seen as a major source of revenue for the government. As we have already seen, the foreign investors are required to pay a 1 per cent import duty on imports of machinery for production, equipment and spare parts. Raw materials and intermediate goods needed for export production are exempt from import taxes. Broadly speaking, there are the following import tariffs:

- 5 per cent for promoted goods such as heavy equipment and machine tools;
- 10 per cent for some medicines and some materials used in light industry, such as fabrics and some chemicals;
- 20 per cent for some food products such as frozen fish; 30 per cent for certain kinds of fruit and vegetables; and 40 per cent for automobiles.

In the case of computers and television sets the import tax is of the order of 20 per cent. At the same time imports of Completely Knocked Down (CKD) and Semi Knocked Down (SKD) kits, with a view to promote local production, attract much less import duty of 5 per cent and 15 per cent respectively.

As a member of AFTA, and with observer status in WTO, Lao PDR is committed to do away with all the restrictions on trade in commodities and services. Moreover, with a view to facilitate trade the government is making efforts to reduce the paper work involved in the process of exporting and importing. With a view to promote exports, an Export Promotion Division has been set up in the Ministry of Trade. The Export Promotion Division has active interface with international Export Promotion Organizations like International Trade Center (ITC), Japan External Trade Organization (JETRO) and also with the Export Promotion Department of Thailand, which is a major trading partner of Lao PDR. The Export Promotion Department actively participates in major international trade fairs, holds seminars and exchanges trade delegations.

The trade promotion efforts by the Export Promotion Division is being complemented by the Lao National Chamber of Commerce and Industry (LNCCI), a non-profit organization set up in 1989 through a decree issued by the prime minister's office. Among other things, the LNCCI acts as an intermediary between the government and the business community, makes recommendations to government on various policy matters relating to trade and industry and also disseminates government policies to business community. For export promotion, LNCCI undertakes various initiatives which *inter alia* include, developing a data base, including the compilation of pertinent trade information within the country and abroad, participation and organization of training courses, seminars etc., on subjects relevant to trade and investment. It also acts as a coordinator between the business community and government authorities, participates in the trade fairs and exhibitions within the country and abroad in order to respond to the needs of members and non-members.

Trade performance

We have made use of the UNCTAD data to present a broad picture of trade performance in the recent past. First, the estimated index of revealed comparative advantage shows that at present the comparative advantage of Lao PDR is confined to just four commodities, namely wood products, clothing, leather products and fresh food. Such a narrow commodity base as well as a high level of product concentration is likely to make Lao PDR highly vulnerable to developments in the markets for these commodities. Secondly, in terms of the direction of trade it is observed that bulk of the exports and imports are with Thailand. This again is not a highly appreciable situation because such high level of regional concentration will make the fortunes of Lao PDR dependent on one country. Naturally, any adverse impact on the major trading partner will get reflected in the export performance as happened during the 1997 crisis.

Under these conditions, the export promotion strategy of Lao PDR might focus on product diversification as well as diversification in terms of trading partners. Given the fact that as of now the comparative advantage is in agro-based industries and those related to leather and textile products, it is important to modernize these sectors. For example, increased productivity in the agricultural sector lies at the root of developing international competitiveness in the agro-based industries. In the wood-based industries and in garments there is greater role for incorporating new technologies like ICT for design development. Thus for the traditional industries to become competitive there is a need for technological upgradation and harnessing the opportunities offered by the new technology. This calls for strengthening the NSI for training of workers, development of design centers and building up of needed infrastructure. In addition to new product development, ICT could also play a critical role in developing new markets through e-commerce. This is more so in the case of service sectors like tourism, which as of now brings substantial returns to Lao PDR. However, going by the available evidence e-commerce is still in its infancy in the Lao PDR.

A crucial issue is whether the production of IT goods offer any opportunity for Lao PDR in diversifying the export basket. Given the land-locked nature of the country, engaging in the production of any transport-intensive goods may not be advisable to the country. In this context, as we have already shown, it might be plausible for Lao PDR to enter into the production of certain ICT goods with low level of techno-logy like passive components, electromechanical components and other intermediate goods for the ICT industry and less skill-intensive IT-enabled services. Here again, trade and investment liberalization alone may not have the desired outcome unless institutional interventions and policy measures are initiated towards developing a vibrant innovation system.

IV. IT production and use

As of now there is hardly any IT production base in the country, which in turn leads to a situation wherein the entire domestic demand is being met through imports.[17] This has had the effect of adversely affecting the overall trade balance of the country on the one hand and forgoing the potential opportunities for employment and income generation in the country through ICT production on the other.

ICT use: Telephones

The government initiatives appear to have paid rich dividend in terms of unprecedented growth in the fixed telephone lines in the country. The number of fixed telephone lines increased from 7270 in 1991 to 47,810 in 2001, recording an annual growth rate of over 20 per cent, the 3rd highest growth rate in the Southeast Asian region (ITU 2003). As a result there has been a more than fivefold increase in the fixed telephone density (0.17 per 1000 people in 1991 to 0.89 in 2002).

The experience of most countries might suggest that "growth breeds inequality", at least in certain sectors. This appears to have also been the case of Lao telecom sector. Impressive growth in the telecom density notwithstanding, there appears to be a marked inter-regional variation in the access to telecom services in the country. The point may be further elaborated: in 2002, it was observed that the Vientiane municipality, which has only about 11 per cent of the total population, accounted for about 65 per cent of the total telephone lines. The telephone density of Vientiane municipality thus turned out to be 3.4 per 1000 people keeping it head and shoulders above other provinces. On the other end of the "intra-national telecom divide" were provinces like Phingsaly wherein the telephone density was found to be as low as 0.022 and the telephone density for the country as a whole (0.606) was only about one-sixth of that of Vientiane municipality (Table 5.4). Hence one of the major challenges ahead appears to find ways and means by which the present "intra-national telecom divide" could be mitigated, if not bridged altogether, by providing telecom access to a large mass of rural population in the country.

Table 5.4 Inter-regional variations in the access to telecom services in Lao PDR

Province	Line capacity	Number of subscribers	(%) of total	Mid-year estimate of population for the year 2000	Number of subscribers per thousand population
Phongsaly	0	38	0.12	1,744,000	0.022
Luang Namtha	1,000	446	1.41	1,309,000	0.341
Bokeo	512	50	0.16	1,296,000	0.039
Oudomxay	1,000	504	1.59	2,398,000	0.210
Sayaboury	1,024	112	0.35	3,328,000	0.034
Luang Prabang	2,064	1,573	4.97	4,161,000	0.378
Houaphan	512	349	1.10	2,791,000	0.125
Xieng Khoung	1,000	506	1.60	2,288,000	0.221
Vientiane Sayssomboune	1,024	361	1.14	3,269,000 6,17,000	0.110
Vientiane Municipality	26,751	20,628	65.21	5,978,000	3.451
Bolikhamsay	1,048	742	2.35	1,866,000	0.398
Khammouane	2,392	1,280	4.05	3,108,000	0.412
Savannakhet	3,880	2,635	8.33	7,662,000	0.344
Saravan	696	389	1.23	2,923,000	0.133
Sekong	0	23	0.07	732,000	0.031
Champassak	2,696	1,943	6.14	5,719,000	0.340
Attapeu	0	53	0.17	994,000	0.053
Total	45,599	31,632	100.00	52,183,000	0.606

Sources: UNDP (2002) and the National Statistical Center (2002).

Telecommunication: Mobile telephones

In a country like Lao PDR, where "last mile connectivity" is a major issue, mobile technology has often been considered an alternative technology to address the telecom needs. Thanks to the various initiatives made by the government, there has been an expansion of mobile network in the country. As of now, although 50 per cent of the provinces in the country have been covered by mobile network (GSM technology), the real coverage, as in the case of fixed telephone, is limited to urban areas covering about a 10 per cent of the population. Lower coverage notwithstanding, the number of mobile subscribers increased from 300 in 1993, the year in which cellular service had its beginning, to nearly 30,000 subscribers in 2001 (Table 5.5).

The observed performance in mobile telephones appears to be much desired when compared with the performance of a comparable neighboring country like Cambodia where the mobile diffusion rate has been more than twice that observed in Lao PDR. More importantly, the access to mobile services is found to be more regionally skewed than in fixed telephones. To illustrate, it has been found that about 80 per cent of the mobile subscribers belong to Vientiane Municipality that accounts for only about 11 per cent of the total population. The next closest figures to this are Savannakhet and Pakse (at 5.8 and 5.6 per cent) respectively. Thus three urban centers put together account for more than 90 per cent of mobile services in the country. Hence, as in the case of fixed telephones, the task before the government appears to be to devise appropriate enabling policy measures for promoting and facilitating greater

Table 5.5 Trends in mobile cellular penetration in Lao PDR

Year	Number of subscribers (000)	Mobile density (per 1000 people)
1993	0.3	0.01
1994	0.6	0.01
1995	1.5	0.03
1996	3.8	0.08
1997	4.7	0.09
1998	6.4	0.12
1999	9.4	0.18
2000	13.9	0.27
2001	29.5	0.52

Source: UNDP (2002).

investment in such a way that the mobile capacity is enhanced through competitive environment.

Internet

Lao PDR is one the last countries in the Southeast Asia to take initiatives to provide Internet access to its people. The delayed entry has been partly attributed to the government's suspicion about the destabilizing influence of the Internet (ITU 2003). But today the government recognizes the fact that while the Internet has some negative influence these are more than offset by its positive contribution towards the economy and society. This has been evident from the government initiatives discussed above. As of now, there are three ISPs in the country[18] (Laotel, Globenet and PlanetOnline), where Laotel is the market leader accounting for about three quarters of the market. Latest available estimate on the number of Internet users indicate that there are about 2900 subscribers.[19] Of this about 47 per cent was accounted by foreigners, 23 per cent by local people, 20 per cent by the local companies and 10 per cent by the government (Phissamay 2002). Table 5.6 indicates the dramatic increase in the number of subscribers and the number of Internet cafes, *albeit* from a rather low base during a short span of one year. Also, the table indicates a marked decline in the cost of Internet service fees. While these developments need to be seen in the context of a pent-up demand on account of the delayed entry of Lao PDR into the Internet world, the role of state initiatives cannot be underplayed.

Promoting IT use: The challenges ahead

Fairly impressive quantitative performance notwithstanding, there are real challenges before the Lao Government in terms of harnessing the new technology for development. This is because the access to

Table 5.6   Indicators of Internet development in Lao PDR

Indicators	2000	2001
ISPs (number)	2	3 (3 more in the offing)
Internet café (number)	11	60
Internet subscribers (number)	1,934	2,900
ISP service fees	$2/hour	$0.5/hour
Café service fees	$3/hour	$0.6/hour
User expenses	$60/month	$ 30/month

Source: Phissamay (2002).

Internet as well as telecommunication is confined mainly to the urban centers, and the rural areas that accommodate bulk of the population remain beyond the ambit of the new technology. Landline and cellular telecom systems work well in metropolitan areas and smaller cities where subscribers are located in dense clusters that justify the high cost of installation. However, connecting rural areas is a bigger challenge because subscribers are geographically dispersed, sparsely populated and economically weak. Telecom companies, in general, may not venture into remote villages because the purchasing power in these villages is not enough to recover the cost of connecting them. Therefore, affordability, ease of deployment and appropriate organizational innovations are critical to sustainable deployment of telecom systems in rural areas. These issues could be addressed through technological and organizational means. For example, adoption of innovations in Wi-Fi (wireless fidelity) technology could take developing countries a long distance in addressing the last mile connectivity. A notable pilot project to incorporate Wi-Fi technology in India is the Digital Gangetic Plain pursued at IIT Kanpur as part of Media Lab Asia program. This project has established a multi-hop Wi-Fi network across 75 kilometers between the Indian cities of Kanpur and Lucknow. It offers the potential to provide a proof-point for enabling rural communications for both voice and data. Another technology project that has gone on to the pilot stage is DakNet in the Indian state of Karnataka, where it is attempting to extend the Bhoomi Land Records project to kiosks that do not have connectivity. DakNet offers Wi-Fi-based asynchronous broadband linkage where wired communication is not available.[20]

The issue of affordability arises mainly on account of the high price of hardware and software in relation to average income level of people in countries like Lao PDR where annual per capita income is only around $350. The developments in the FOSS, though in its early stages of development, are likely to provide an alternative for countries like Lao PDR. Jhai Foundation[21] is localizing the Linux-based graphical desktop and office tools to the Lao language. Several Indian groups are actively at work localizing FOSS to Indian languages, and these include groups like Malayalam Linux, Tamil Linux and others. FOSS, therefore, is increasingly becoming a relevant alternative for developing countries like Lao PDR.

In addressing the affordability and access issues, there are a series of institutional intervention in countries like India as well as in neighboring Thailand.[22] These attempts are being supplemented by the development of low-cost devices. Several organizations like the Jhai

Foundation in Lao PDR and the Simputer Trust[23] in India are innovating computer technology to make it more appropriate for rural areas. The Jhai PC, for example, gets around the lack of power in rural areas by running low-wattage computers that are driven by pedal-power and that can survive harsh rural conditions. The Simputer Trust is building a multi-user computer that will be a shared device which can be personalized through the use of smart cards issued to individual users. These are early examples of applying ICT for development, particularly in rural areas, and are therefore of much relevance for Lao PDR.

V. Concluding observations

In an economy wherein almost 80 per cent of the population depends on agriculture and allied rural sectors, a major challenge lies in harnessing the new technology for addressing the development issues of the rural sector. It is evident that the government of Lao PDR is highly committed towards addressing this issue and heroic initiatives have been made especially during the last decade. Nonetheless, given the gigantic task at hand and the rocky road through which it has to traverse, the destination still remains far away. Thus there is a challenge and also an opportunity. Our inquiry tends to suggest that IT use in Lao PDR, viewed in terms of telecommunication network (fixed or mobile) and Internet use is confined to the urban areas leading to what is called the "intra-national digital divide".

To address this issue a two-pronged action may be worth considering: first, a concerted effort to build up an innovation system with focus on human capital which is important not only for generating needed local content but also for enabling people to make use of the new technology. With regard to the generation of human capital stock, the study calls for (a) an accreditation scheme for the different institutes, (b) reviewing the present approach of "training the trainers" because this approach is based on the basic premise that all those who are trained will be willing and able to train others, (c) critically evaluating the present "top-to-down" approach towards IT in education and exploring the possibility of a "bottom-up" approach, and (d) creating more training facilities within the country, in association with those renowned universities/institutes in the neighboring countries. Secondly, attracting more investment into the field of IT production such that new income-earning and employment opportunities are generated. More investment is also called for in IT infrastructure to ensure a more competitive environment such that the costs and prices of telecom services are brought down.

To the extent that the intra-national digital divide is a manifestation of the low affordability on account of the low purchasing power and higher price of computer hardware and software in relation to the per capita income in the country, the study calls for collective efforts in collaboration with other less developed countries with more IT capability like India. Given the fact that the IT production base in the country at present is negligible, there appears to be the need to develop an IT production base in the country so that IT use is facilitated and new avenues of employment and income are created.

The study finds that the series of initiatives made by the government towards attracting more investment into the country notwithstanding, there appears to be ample scope for further improvement. For example, the study tends to suggest that the stipulation of minimum capital requirement of $0.1 million for the fully owned foreign enterprises is likely to have the effect of erecting entry barriers to certain foreign enterprises, especially the small- and medium-sized ones. Similarly, setting the minimum foreign share of 30 per cent for the joint ventures also might act as an entry barrier.

The study tends to suggest that the domestic investment tends to crowd in foreign investment. Since the domestic private sector, for historical reasons, is in its infancy the study calls for further targeted efforts to develop a domestic private sector. This process could start by reviewing and abolishing policies that are present today, which discriminate the local capital *vis a vis* their foreign counterparts. As already mentioned, special emphasis may be given to create and nurture an entrepreneurial class in the country through EDPs.

In the field of IT the private investment today is confined to the telecom services and IT training. Though there are limits set by the international environment in which IT production is being organized, the study tends to suggest that Lao PDR could consider entering into the production of low technology IT goods like IT entertainment goods, IT components like passive and electro-mechanical components and IT-enabled services. However, given the global production network of ICT, there is an inexorable link between trade and investment in IT production. Therefore, to facilitate investment in IT production it is important to ensure a conducive trade regime.

While venturing into such initiatives, it is important to keep in mind the lessons offered by the experience of other countries. To begin with, the strategy needs to be not one of spreading thinly the resources across the country, instead the investment needs to be undertaken in such a way as to take advantage of the agglomeration economies. This might be possible through the setting up of technology parks wherein built-up space, communication infrastructure and others, which are beyond the reach of an individual entrepreneur, are provided along with a "single window clearance" system so that the prospective investors need to have only limited interaction with the bureaucracy.

Secondly, such technology parks need to be close to and have constant interaction with the centers of learning such that mutual learning and domestic technological capability is built up in the long run. Thirdly, there is also the need for conscious efforts towards building up an innovation system with focus on skill empowerment such that the economy does not get locked up in low technology activity, and an upward movement along the skill spectrum is ensured. It needs to be noted that on-the-job training has already been emphasized in the Lao PDR policy towards investment. However, given the fact that skill empowerment is bound to bring about substantial social benefit, its cost cannot be left entirely to the private sector and hence it may be shared socially. This could be accomplished, perhaps, through providing additional incentives for firms undertaking such initiatives.

6
Myanmar: Sowing but Not Harvesting?

I. Introduction

Myanmar presents the case of an economy that makes earnest efforts towards regaining the glorious past through a series of market-oriented reforms initiated since 1988. But the structural transformation of the economy being pursued through policy reforms is yet to be completed; in 1936, prior to independence, the share of agriculture in GDP was of the order of 66 per cent and the situation seemed to have not changed drastically even in 2002. Going by the data provided by World Bank (2002)[1] on the contribution of GDP by different sectors of the economy over time, a critic might present the case of Myanmar as a case of "structural retrogression", and contrary to the experience of most developing countries.

The share of primary sector increased from 47 per cent in 1981 to nearly 59 per cent in 1991 and recorded a marginal decline in the following decade to reach a level of 57 per cent in 2000, whereas most developing countries have witnessed decline in the share of primary sector with a corresponding increase in the share of other two sectors. Needless to say, while the share of agricultural sector increased, that of industrial sector and services recorded a corresponding decline. To elaborate further, the share of industrial sector declined from a little over 12 per cent in 1980 to 9.7 per cent in 2000. Within the industrial sector, the share of manufacturing sector declined by about 2 per cent during the period under consideration from about 9 per cent in initial year (1980). The share of service sector also declined from about 40 per cent in 1980 to 33 per cent in 2000. This however, is not to neglect the marginal increase of about 2 per cent recorded in the share of both the industrial and the service sectors during 1990–2000. If the above evidence is any

139

indication, Myanmar appears to be presenting an unconventional development trajectory, wherein not only the primary sector dominates the share of industry, more specifically manufacturing sector and services recorded near stagnancy in their share in GDP.

At the same time, it must be noted that the government has been highly committed towards bringing about all-round growth and transformation of the economy and welfare of people at large. It is evident from the economic objectives upheld by the government (Box 6.1).[2]

Box 6.1   Economic objectives upheld by the Government of Myanmar

Economic objectives

- development of agriculture as the base and all-round development of other sectors of the economy as well;
- proper evolution of the market-oriented economic system;
- development of the economy inviting participation in terms of technical know-how and investments from sources inside the country and abroad;
- the initiative to shape the national economy must be kept in the hands of the state and the national peoples.

In achieving the economic objectives the government has also been taking a series of initiatives for attracting investment with more freedom for private sector. The per capita investment in the economy has more than doubled from about 233 to 586 Kyats during the last 15 years (1985–86 to 2000–2001) recording an ACGR of 6.34 per cent (at 1985–86 prices). The recorded ACGR in total investment was still higher – almost 8.5 per cent during the period under consideration. The crucial issue, however, is how efficient this investment has been or what has been the rate of return to the increased investment in terms of output growth? Higher growth in investment notwithstanding, the net output per worker, an indicator of labor productivity, has recorded a negligible growth rate of 2.07 per cent during 1985–86 to 2000–2001. More importantly, given the lower level of productivity, the growth in GDP seems to have failed to keep pace with the population growth rate and the net result has been a negligible growth (ACGR) in per capita GDP at about 1.88 per cent. No wonder, the per capita consumption declined from about 1336 Kyats in 1985–86 to 1281 Kyats in 2000–2001.[3]

The evidence presented above tends to suggest that for an economy like Myanmar, which is richly blessed with natural and human resources to achieve the declared objectives, there is the imperative of, among other things, increasing the rate of investment on the one hand and improving the overall level of efficiency on the other. In a context wherein the budget deficit is growing in leaps and bounds (from about 11,204 million Kyats in 1990–91 to 1,09,725 million Kyats in 2000–2001 – almost tenfold increase),[4] there are obvious limits for the government to fill the investment gap. Hence, there are not many options open, but to continue with the ongoing liberalized policy towards domestic and foreign investment and doing away with constraints if any that remain. To address the issue of efficiency, productivity and international competitiveness, there appears to be the need for an overall technological transformation, which in turn calls for building up an innovation system with increased access to both embodied and disembodied technology coupled with bringing about a more competitive environment. At the same time, given the all-pervasive nature of IT and its capability to bring about efficiency, productivity and competitiveness in almost all sectors of the economy, the role of IT in bringing about the structural transformation cannot be overemphasized. Here lies the need for trade and investment reforms and building up a NSI – the central issue being addressed by the present study. Being a founding member of WTO, and new entrant to ASEAN and signatory to e-ASEAN Framework Agreement, some of these issues have been getting reflected in the series of policy reforms initiated by the government since 1988.

II. NIS in the IT sector of Myanmar

Policy measures

Though the government did not have an explicit IT policy till 2003, significant initiatives have been made by the government not only for promoting the production, but also for harnessing the new technology for the growth and development of the economy and society. For example, the government passed the Computer Science Law as early as in 1996. In 2002, the Myanmar Computer Science Development Council[5] developed a draft IT Master Plan for the country. The Master Plan explicitly dealt with all the different aspects of IT production and use and the MPTC issued additional guidelines to govern the use of computers and Internet in the country. What follows is a brief examination of these policy documents to reflect on likely policy options for furthering the production and use of ICT in the country.

The Computer Science Law (1996)

The use of computers in the country has been guided by the Computer Science Development Law passed by the Government in 1996.[6] The law aimed at promoting computer education in the country and building up IT manpower base to promote the use of computers and to contribute towards the emergence of a modern developed state. The law had the objective to supervise the import and export of computer software and information.

The law also formed the Computer Science Development Council (CSDC)[7] which has the following duties and powers:

- laying down the policy and giving guidance for the development of computer science in the state to keep abreast with the times;
- laying down the policy with respect to the systematic dissemination of utilization of computer science in the state;
- laying down the policy, giving guidance and controlling with respect to computer network;
- making arrangements for the youth, especially students to get the opportunity of studying basic computer science;
- laying down the policy, giving guidance and controlling with respect to IT;
- supervising and giving guidance with respect to activities of the federation and computer-related associations formed under this Law;
- prescribing the types of computer software and information which are not permitted to be imported or exported;
- laying down measures to cause extensive development in the utilization of computer science in the respective fields of work in the state;
- forming necessary working committees and bodies related to computer science and assigning duties thereto;
- abolishing any computer association formed or existing not confirming with the provisions of this law or any computer association not functioning in conformity with the provisions of this law or not in conformity with the constitution of the relevant association;
- laying down and carrying out measures necessary for the attainment of the objectives of this Law.

The law also made the provision for and laid down the guidelines for the formation of computer associations and the formation of the Federation, comprising representatives chosen from the Myanmar Computer Enthusiasts Associations and Myanmar Computer Entrepreneurs Association. The law also laid down the duties and powers of the Federation, which deal with different aspects of IT manpower generation, development of an IT industry and promoting the use of computers (for details see Box 6.2).

Box 6.2 Duties and powers of the federation in Myanmar

(a) carrying out research for the development of computer science in the State to keep abreast with the times;

(b) conducting research in computer science, giving assistance to the persons conducting research;

(c) promoting extensive utilization of computer science in the respective fields of work;

(d) prescribing the syllabi and curricula for computer training schools;

(e) inspecting teaching in computer training schools as may be necessary so as to determine whether it is up to the standard or not;

(f) running computer science courses, holding lectures, competitions and organizing study tours;

(g) holding examinations in computer science, conferring certificates and medals;

(h) submitting advice to the Council from time to time on the development of computer science;

(i) giving assistance to manufacturers so as to enhance the quality of computer hardware and software;

(j) giving assistance for production of computer hardware and software and for sale inside and outside the country;

(k) laying down projects on IT in accordance with the guidance of the Council;

(l) communicating with international computer organizations;

(m) making arrangements for holding and dispatching delegates to local and foreign conferences, meetings, workshops, seminars, paper-reading sessions as may be necessary;

Box 6.2   (Continued)

(n) fulfilling a target to devise a system that can use Myanmar
     language in the computer;
(o) tendering advice to government departments and organiza-
     tion which seek advice with respect to computer;
(p) compiling, publishing and distributing books, papers, period-
     icals and journals on computer;
(q) setting up a library to collect books on computer from inside
     and outside the country;
(r) carrying out training for the youth, especially students, to
     acquire basic computer knowledge and to facilitate cause
     emergence of outstanding computer scientists;
(s) awarding monetary prize to outstanding computer scientists
     and inventors;
(t) recommending to the Council to confer honorary titles and
     awards on outstanding computer scientists and inventors by
     the state;
(u) submitting advice to the Council in order to protect the bene-
     fits of computer scientists and inventors;
(v) forming necessary committees and bodies, and determining
     the functions and duties of those;
(w) carrying out tasks with respect to computer science assigned
     by the Council.

It is also laid down by the law that

Whoever imports or keeps in possession or utilizes any type of
computer prescribed under sub-section (a) of section 26, without the
prior sanction of the Ministry of Communications, Posts and Telegraphs shall, on conviction be punished with imprisonment for a
term which may extend from a minimum of 7 years to a maximum
of 15 years and may also be liable to a fine. (p. 6)

Also

Whoever sets up a computer network or connects a link inside
the computer network, without the prior sanction of the Ministry
of Communications, Posts and Telegraphs shall, on conviction be

punished with imprisonment for a term which may extend from a minimum of 7 years to a maximum of 15 years and may also be liable to a fine. (p. 6)

Later, a number of laws to regulate different aspects of Internet use were issued by the Myanmar Posts and Telecommunications (MPT). Given the fact that free flow of information is a *sine qua non* for harnessing the new technology for development, the series of restrictions are likely to undermine the effect of various initiatives that the government has undertaken to use IT for development. Hence, phasing out, if not abolishing altogether, of restrictions on the use of Internet and computers seems to be a promising policy option.

The IT Master Plan

The IT Master Plan, which has been prepared by CSDC based on a systematic analysis of the present status and future trend of IT in Myanmar and rest of the world, specifies the mission, strategies and action plans for the period up to year 2010. The Master Plan rightly foresees many opportunities for countries like Myanmar. While discussing the status of IT in Myanmar, it recognizes that the IT level of Myanmar is still at the early stage in comparison with the international standards. Hence calls for considerable effort to develop IT in the country wherein the state has been assigned a key role by doing the tasks which cannot be done by private sector such as building IT infrastructure, amalgamation of law, rules and regulations, setting standards and then creating an environment for fair competition. At the same time, the Master Plan also calls for, among other things:

- coordination and cooperation between public and private sectors;
- improving the quality of IT manpower through effective human resource management;
- improving the IT awareness among the responsible personnel at ministry, directorate and private enterprise levels;
- overcoming the shortage of books, journals, magazines teaching and learning materials for teachers and trainers;
- taking IT beyond cities like Rangoon and Mandalay;
- identifying the competitive factors and earnest effort to attain them.

The Master Plan highlighted the need for widespread application of IT in the state management with the intention of providing better services to the public; improving efficiency and reducing costs in the business organizations to improve productivity and efficiency. It called for creating an IT intelligent society, reducing digital divide, facilitating e-commerce and improving the educational level. The Master Plan also underscored the need to develop an IT industry and to develop human resources to facilitate the production and use of IT.

The strategy underlines *inter alia* the need for public–private cooperation, creation of demand base by the state, provision of incentives for business organizations to promote IT use, foreign investment and technology transfer, creating a software industry, promoting international cooperation, provision of Internet access, establishing an IT zone, facilitating a liberalized investment climate and developing the legislative framework for the promotion of e-commerce.

For the development of IT in Myanmar by the year 2010, the following priority areas have been identified:

- IT application
- IT in Education
- foundation for IT Industry
- ICT Infrastructure and
- legal Infrastructure.

IT application essentially implies the development of information systems for public organizations which come under the rubric of e-governance and include, but not limited to, computerization of public organizations with a view to achieve cost reduction, efficiency enhancement and greater access to information by the people at large. It also includes creation of better database on all economic activities including science and technology, natural resources and environment, information system for the health care system and use of IT for enhancing international competitiveness. Here the Master Plan calls for appropriate incentive systems to encourage application of computers.

IT in education involves the development of a human resource for IT at the national level, certification scheme for computer professionals, provision of computers in all the courses of study, increased Internet access in a phased manner (beginning with universities and colleges, later for the high schools, middle and primary schools), program for providing IT education for government employees, implementing a scheme of training the IT trainers and so on.

Creating a foundation for IT industry is envisaged through generating an adequate supply of software engineers for the establishment of software industry, provision of government support for entering the international market and by establishing an IT zone wherein infrastructural facilities and single window clearance system for facilitating investment.

Construction of IT infrastructure involves building up a national data communication network connecting local and international networks. Given the high cost of the project, and the critical role of IT in business and for developing the IT production base, the Master Plan proposes to provide a data network in the first phase for providing communication facilities for the business organizations and IT industry. Further the plan underlines the need for integrating the information systems of public and private organizations and calls for a standardization committee.

Finally, the plan calls for the development of regulatory and legal framework that creates IT confidence for consumers and facilitates the wider use of IT. This involves the formulation of laws and policies like the laws for e-commerce, digital signature, electronic transactions and payment settlements at national and international levels, protection of intellectual property, establishment of authentication authorities and so on.

By realizing the fast-changing nature of technology and the inherent difficulties involved in predicting the likely changes, the plan also calls for appropriate changes, whenever necessary, so that the plan can be successfully implemented.

IT infrastructure: Radio and television

In an economy wherein almost a little over 75 per cent[8] of the population live in rural areas, it is natural to expect greater relevance for TV, radio and newspapers. According to the data provided by the World Bank (2004) the number of radios per 1000 people in Myanmar is found to be 66. This has to be compared with 139 in the low-income countries and 112 in South Asia in 2001. Also, it has been observed that in 2001 there were only 11 countries reporting lesser radio density than Myanmar. Relatively higher level of literacy notwithstanding, number of daily newspapers is also found to be only 9 per 1000 people, which is also at a rather low level.

But in the case of television, as a result of the concerted effort by the government, TV transmission now reaches the entire country, including the remote areas. As of 2000–2001, there are over 101 TV relay stations in the country and it may be noted that, except Magway division, all the states and Divisions are having at least one TV relay station. The highest number of TV relay stations is reported in Shan state (45) followed by Kachin (15) and Sagaing Diviosn (9).[9] The achievement becomes more striking when we consider the fact that in 1990–91 there were only five relay stations in the country and the recorded increase has taken place in a short span of five years (during 1995–96 to 2000–2001). It has been reported that at present television coverage is over 82 per cent or TV broadcast can be received in 267 out of 324 townships of Myanmar.

The commendable efforts by the government towards increasing the TV transmission coverage notwithstanding, the number of TV sets is found to be only 8 per 1000 people and this has to be compared with 91 for the low-income countries in general and 84 for South Asian countries. Also in the year 2002, as per the data published by the World Bank (2004), there were only ten countries reporting a lower TV density than Myanmar. The cable TV is yet to begin in the country. Hence, home satellite receivers are found very popular at least in the capital city. Their number has almost doubled during the six years beginning with 1994–95 (Table 6.1).

Low level of diffusion of TV and radio, notwithstanding the earnest efforts towards the building up of needed infrastructure by the state, tend to suggest that the infrastructure and public investment *per se* may not be instrumental in taking the technology to the people. There are indeed a number of complementarities to be ensured if the huge investment by government is to provide corresponding returns. After all, "railroads without locomotives make no economic sense". It appears that such complementarities are not adequately available in Myanmar. To begin with, people need to have the purchasing power. In a country

Table 6.1   Number of licenses issued for radios, TVs and satellite receivers and the revenue collected in Myanmar

Year	No. of radio licences	Revenue (million kyats)	No. of TV licences	Revenue (million kyats)	No. of home satellite licences	Revenue (million kyats)
1990–91	3,916	0.12	129,036	15.59		
1994–95	71,271	1.89	223,886	28.6	995	5.51
1995–96	22,643	0.81	279,251	35.4	1,716	32.33
1996–97	13,007	0.41	282,504	33.93	1,823	24.96
1997–98	32,293	1.11	284,642	35.4	1,754	20.4
1998–99	35,591	0.38	260,724	32.54	1,411	18.49
1999–2000	9,380	0.15	278,161	34.77	1,679	25.42
2000–01	18,500	0.27	396,353	31.78	1,758	29.47

Source: Central Statistical Office (2001).

wherein the per capita income is only about $300 (World Bank 2002) and consumption has shown a declining trend coupled with galloping inflation (ADB 2004), the rate of diffusion of the so-called "luxury items" like television is at a low level. The low level of affordability is further compounded by the licensing system for radios, TVs, satellite receivers, and VCRs. It is also to be noted that the revenue from radio license is not substantial and has been showing a declining trend (see Table 6.1). Given the importance of radio and TV for the rural masses it may be worth considering the abolition of the licensing system. If revenue is the prime consideration it may be worth considering imposing a lump sum as license fee at the time of purchase. Such a strategy might also be instrumental in bringing down the transaction cost involved in license fee collection, both by the Government and by the owners of these gadgets.

Telecommunication

Telecommunications were introduced in Myanmar in 1979 when a satellite communication ground station was commissioned at Thanlyn, across the Bago River from Yangon (Rangoon), and microwave systems were installed in Yangon and Mandalay. As the government considers telecom as a priority area for development, and it being an integral part of IT, there has been series of initiatives towards building up a modern telecom infrastructure. The result has been a continuous upgradation of the telecommunication system at a rather fast rate with additional ground stations, microwave connections and digital exchanges provided by China, Singapore, Israel, Japan, Germany and Australia.[10] The Install-ation of 960-microchannel digital microwave links at the Thanlyin Satel-lite Communication Station (Phase I) has resulted in further improve-ment in the telecom system. On the whole, during the last 13 years, modern digital auto exchanges have replaced the old communication system and a most modern cellular phone system (CDMA – the Code Division Multiple Access) has also been established.

Computers and Internet

We have already seen that Internet in the country has been subjected to a series of restrictions. This stands in the way of taking advantage of the full potential of IT for development and therefore calls for ways and means of at least phasing out some of these restrictions. Against this background, the government launched Myanmar intranet in 2001. Besides members at Bagan Cybertech, MPT mail users also can view the local intranet. Internet users abroad can also access Myanmar intranet.

There are four types of users for the intranet[11]

1. *Member A* – has the access to use File Transfer-Protocol (FTP) service, Intranet Service, and dial-up web access (web browsing). The user can request sites that he/she wants to see. The user can use this service from home or office.
2. *Member B* – The services provided are the same as Member A, but has to use it at the main office of Bagan Cybertech. FTP service is not provided.
3. *Member C* – provided FTP service only.
4. *Member E* – provided Intranet service only.

With a view to provide the necessary communication infrastructure for the IT companies, Bagan Cybertech was established in October 2000 as an economic enterprise. Bagan Cybertech has also built the Bagan Teleport, which has the infrastructure for using the biggest wireless network in Myanmar. VSAT Network uses TDMA technology (Time Division Multiple Access), which is the latest in satellite communications. About 50 terminals in Yangon were using VSAT Network and it has the capacity to connect up to 200 terminals. The cost of a VSAT unit is US$20,000 and the monthly fee is US$250. The service was initially started in Yangon but later extended to Mandalay.[12]

The publicly owned Bagan Cybertech has doubled its available broadband delivery speed from 256 to 512 kbps in early April 2003. The activation fee for broadband wireless access is K1.95 million for an individual account and K2.2 million for corporate subscribers. According to the speed of the connection and the volume of data downloaded, the monthly fee for individual users ranges from K28,000 to K76,000 and those for the corporate users range from K1,20,000 to K2,00,000.[13] The monthly fee for an optional telephone is fixed at K16,000 for individuals and K24,000 for corporate users. This is indeed relatively low when compared to the cost of fixed line (K5,00,000) charged by MPT.

To promote software development and export from the country, an STP has been set up and another one is being planned. The Park has been jointly set up by the government and the private sector, has incubation facilities and envisages an active linkage with the universities.

Human resource development in IT

Realizing the importance of higher education and skill building, especially in high technology areas, the government has taken proactive steps in promoting education in science and technology in general and IT in particular. Major institutions of higher learning have been kept under the administrative control of the S&T Ministry to have better focus and attention. These are

(a) Yangon University of Technology
(b) Mandalay University of Technology
(c) Pyay Technology University
(d) Yangon University of computer studies and technology
(e) Mandalay University of computer studies and technology.

It could be observed that out of the five institutions two are specialized universities focusing exclusively on IT (d and e). The Yangon University of Computer Science and Technology, the leading university in IT education, provides 12 courses in IT education with varying duration (Table 6.2).

In addition to these universities, there are 24 government colleges and 80 university colleges and all of them have IT departments and offer diplomas or degrees in IT education. All these universities are having access to computers with Local Area Network (LAN). It is understood from discussion with the senior faculty of universities under the Ministry of S&T that most of them have access to faculty exchange programs and also collaborate with foreign universities. The outturn of students with

Table 6.2 Courses offered at the University of Computer Science

Name of course	Duration
Bachelor of Computer Science	5 years
Bachelor Computer Technology	5 years
Master of Computer Science	3 years
Master of Information Sciences	3 years
Master of Computer Technology	3 years
Diploma in Computer Sciences (part time)	2 years
Diploma in Computer Science (full time)	1 year
Diploma in Computer Science (full time, Ministers office)	1 year
Diploma in Computer Science (full time, Defence)	I year
Diploma in Computer Science (full time, USDA Hmawbe)	I year
Computer Maintenance (staff of the Ministry of S&T)	6 months
Computer Maintenance (staff of the Ministry of Defence)	6 months

Source: Kyaw Aye (2002).

IT qualification in 2003 has been 3000 per annum and the government has the target of reaching 25,000 by 2010.

The government also made initiatives to promote IT collaboration with Japan, India and Singapore. To address the issue of IT illiteracy among the government officials and familiarizing them with issues in e-governance, the MPTC organizes e-government training workshops. It is a part-time program with duration of six weeks. During June 2003, the second workshop was organized. It was attended by 59 trainees from 31 ministries.[14]

On the whole, though the policy documents explicitly call for evolving a vibrant innovation system, key elements of an innovation system, though at its preliminary stage, is present in the IT sector of Myanmar. Nonetheless, to reap the benefits of the efforts made, there appears to be the need for doing away with the series of restrictions that are present in the use of IT in the country. Also a key issue is whether are there are complementary policies, particularly in the sphere of trade and investment, with a view to take advantage of the full potential of manpower being generated.

III. Trade and investment regime

Investment policies

In an economy characterized by low saving rate coupled with low level of technological capability, the role of private investment in general and foreign investment in particular cannot be overemphasized. No wonder the government as early as in 1988, as part of the economic reforms, underscored the need for promoting private sector and assigned an increasing role for FDI. This was manifested in the Union of Myanmar Foreign Investment Law (FIL) of 1988, which provided for a propitious environment for the growth of private investment by encouraging entre-preneurial activity. The FIL was followed by the Private Enterprise Law of 1990 aimed at further increasing the role of private sector in the economy. This law was designed primarily to promote the development of small- and medium-sized firms. This law was followed by the Cottage Industry Law to further the cause of small industries development. In 1994, the Myanmar Citizen Investment Law complemented the existing investment laws.

The main agency for all issues related to investment is the Myanmar Investment Commission (MIC) which was established in order to oversee and administer the FIL. Among other duties, the MIC is

responsible for scrutinizing FDI proposals and for issuing specific investment permits after getting approval from the Trade Council and the
Cabinet.

The specific objectives of the FIL are the following:

- promotion and expansion of export;
- exploitation of natural resources, which require heavy investment;
- acquisition of high technology;
- supporting and assisting production and services involving large
  capital;
- opening up of more employment opportunities;
- development of works which would save energy consumption; and
- regional development.

Needless to say these objectives reflect a realistic understanding of the
basic character and needs of the economy on the one hand and the
commitment of the state towards taking the economy to a higher growth
path on the other. At the same time, there appears to be certain issues
that need to be addressed. For example, the law highlights the need for
acquiring technology, which is an important aspect of building innovation system in developing countries. It needs to be noted that there could
be many different ways for accomplishing this objective. The promotion
of FDI is only one among them and the law rightly calls for facilitating
the same. Technology acquisition, analytically speaking, can take the
form of embodied or disembodied technology. The former involves the
import of capital goods and spares which fall in the realm of trade policy.
At the same time, it needs to be noted that a major source of technology acquisition takes the form of transfer of disembodied technology
mainly through technology licensing. Here, the extent of technological
capability that could be built up and other economic benefits depends
on the terms and conditions of technology transfer in terms of royalty
rate, technical fee, duration of contract and others like restrictions on
exports. The law keeps silence regarding these issues.

Secondly, the law rightly underscores the need for regional development. In an economy like Myanmar, wherein the inter-regional variation in the levels of development is substantial this is a highly laudable
objective. Having said this, it appears that the law is not explicit in
terms of achieving this objective. Hence it might be advisable to divide
the economy into different zones based on definite indicators of development, and the incentive structures may be turned in such a way as
to achieve the objective of balanced regional development. At the same

time, given the fact that the regional governments are better able to assess the regional requirements, it is advisable to explore the possibilities of providing greater role for the regional governments. While this might have the potential danger of competition between regional governments and the consequent adverse effects, the MIC could act as a watchdog to modulate the behavior of different state governments.

In accomplishing the stated objectives, the law provides for the setting up of both joint venture and fully owned foreign subsidiaries. In the case of fully owned foreign subsidiaries, the law stipulates that the minimum foreign investment should be at least US$500,000 in the case of industrial sector and US$300,000 in the case of a service organization. The stipulation of minimum capital requirement of $0.5 million is likely to have the effect of erecting entry barriers to certain foreign enterprises; especially the small- and medium-sized ones. This in turn could have its adverse effect on the total investment inflows. This is because, in the current era of globalization, it has been observed that large multinational firms are not the only source of investment resources, management expertise and technology that is badly needed by the developing world. But there are a large number of SME having the financial and other resources and keen on investing in the developing countries. Hence, by opening the doors of investment for such small and medium firms might be instrumental in attracting more investment to Myanmar.

A joint venture could be between a foreign firm and a local partner (an individual, a private company, a cooperative society or a state-owned enterprise). Here the law stipulates that the minimum foreign contribution should be at least 35 per cent. Here again, if the evidence from other countries is any indication, setting the minimum foreign share might act as an entry barrier. Hence it may be worth considering the economic rationale for setting the minimum contribution at 35 per cent for such a minimum contribution appears to be not a necessary condition for reaping the benefits of foreign investment. Perhaps it may be better to leave to the foreign and local counterparts to determine the sharing of investment. Such an approach is also likely to enhance the investment inflows to Myanmar.

The law also provides for a number of incentives, which include among others:

- exemption from income tax for three consecutive years beginning with the year in which the operation commences and further tax exemption or relief for an appropriate period in case if it is considered beneficial for the state;

- exemption or relief from income tax on profit which is reinvested within one year;
- relief from income tax up to 50 per cent on the profit from export;
- right to pay income tax on behalf of the foreign employees;
- right to deduct R&D expenditure;
- right to accelerate depreciation;
- right to carry forward and set off losses up to three consecutive years from the year loss is sustained;
- exemption or relief from custom duty and other taxes on imported machinery and equipment for use during the construction period;
- right to import raw material for the first three years of commercial production following completion of construction.

These concessions and incentives appear appropriate to attract investment as well as maximizing the linkage from the investment in the short term and long term. At the same time there appears to be an element of ambiguity in some of the provisions, which might in turn lead to a situation of lack of transparency and corrupt practices. To illustrate, regarding the provision to give tax exemption beyond the initial three years, it will be given for "appropriate period" and also "in case the project is considered beneficial to the state". How to decide the appropriate period? What is the criterion for deciding the usefulness of the project? From the investor's point of view, it is only appropriate that these are defined in more clear terms. Another point related to the tax exemption for profits on exports is that regardless of the nature of exports such incentives are offered. This tends to suggest that the law does not differentiate between potato chips and microchips. Given the fact that as the entry barriers vary across different commodities, the incentives offered need to have some relationship with the entry barriers faced by the exporters. If government were to attract more investment into the high technology areas and to promote high-tech exports, there appears to be the need for targeting incentives.

At the same time it is also to be noted that in the list of items considered by the government as having opportunities to make investment, IT goods or services at present do not find a place. Given the fact that the government has been making earnest effort towards developing the human capital stock and communication infrastructure, it is also important that more skill- and human capital-intensive sectors like IT which make use of these endowments find due place in the agenda for investment promotion.

Trend in FDI

Other things remaining the same, the effectiveness of FDI policy in any country may be gauged by examining the trend in foreign investment approvals and actual inflow. Hence let us now examine the amount of foreign investment approved as well as the actual inflow into the country. Table 6.3 presents the data on these variables and are highly revealing.

To begin with, investment approval as well as the number of projects showed an upward trend till 1996–97 wherein the number of projects peaked at 78 and the investment approval reached an all time high of US$2814 million. Both the number of projects and the investment approved showed a downward trend from 1996–97 to reach an all-time low level of 7 projects and $19 million investment in 2001–2002. It is, however, heartening to note that, going by the data available for the first six months of 2002–2003, there appears to be a revival not only in terms of number of projects but also in terms of the amount of investment approved. Nonetheless, no definite conclusion is warranted as to if this indicates the beginning of a revival.

The table also presents data on actual investment obtained from UNCTAD (2002). The broad trend in actual investment follows that of approved investment. Using the approved and actual data (which is available for the period from 1995–96 to 2000–2001) we have estimated the fructification rate and it turned to be as low as 29 per cent. To the extent that this represents the case of contract failure, it is important that the government explore the reasons for the same

Table 6.3   Trend in foreign direct investment in Myanmar: approved and actual (US$ million)

Year	FDI approved	No. of cases	Actual inflow
1990–91	280.57	22	
1994–95	1,351.88	35	126
1995–96	668.17	39	277
1996–97	2,814.25	78	310
1997–98	1,012.91	56	387
1998–99	54.39	10	315
1999–2000	58.15	14	253
2000–2001	217.69	28	123
2001–2002	19.00	7	NA
2002–2003*	47.68	5	NA

Note: * The year 2002–2003 refers to the period April–December.

to squarely address them. It appears *a priori* that the FDI policy, apart from focusing on attracting more investment, should also give equal importance to the issue of investment implementation. This calls for identifying the roadblocks which would include strengthening of innovation system at different levels.

Role of FDI in Myanmar economy

The issue is whether additional efforts need to be undertaken to implement the investment proposals has to be based on a fair understanding of the role of FDI *vis a vis* local investment in different sectors of the economy. Data presented in Table 6.4, which shows the share of foreign and local investment till 2000–2001, may be an appropriate pointer. It is evident that in most of the sectors the bulk of the investment was undertaken by the foreign sector. The share of foreign sector was more than 90 per cent in four sectors, and between 80 and 90 per cent in five sectors. To cut the story short, except the construction sector in all the other sectors foreign investment is predominant.

Now let us examine the revealed preference of the foreign investors in terms of the mode of investment. Given the present law any foreign investor could have the following choices – fully owned foreign subsidiary, joint venture, on product-sharing arrangements. The joint venture could be with local private sector, state economic enterprises, Myanmar Economic Holdings Ltd, Yangon City Development Committee, Cooperative societies and the Myanmar Economic Corporation. For our purpose

Table 6.4 Share of foreign and local investment in different sectors in Myanmar till 2000–2001

Sector	Local investment (%)	Foreign investment (%)
Agriculture	11.07	88.93
Fisheries	17.99	82.01
Mining	5.55	94.45
Oil and Gas	0.00	100.00
Manufacturing	13.89	86.11
Transport	0.12	99.88
Hotel and Tourism	11.41	88.59
Real Estate Development	0.59	99.41
Industrial Estate	39.88	60.12
Construction	93.81	6.19
Others	21.45	78.55
Total	13.91	86.09

Table 6.5   Distribution of foreign investment approval across different ownership categories in Myanmar (%)

Year	Wholly foreign	JV private	JV others	Production sharing basis
1990–91	28.0	–	46.7	25.3
1994–95	5.00	4.06	56.91	34.03
1995–96	67.89	0.28	26.87	4.96
1996–97	38.52	2.05	34.62	24.81
1997–98	51.21	23.32	7.91	17.56
1998–99	44.83	5.75	40.32	9.10
1999–2000	57.13	3.83	2.04	37.00
2000–2001	48.43	11.81	17.64	22.12

Source: CSO (2001).

we have divided them into four categories: foreign subsidiaries, joint ventures with private sector, joint ventures with others and Production-sharing arrangements. Table 6.5 presents data on the share of each of these categories during the last decade. Though no clear trend emerges from the table, inferences, though tentative, may be made. To begin with, the most preferred form of investment appears to be setting up fully owned foreign subsidiaries. Secondly, there appears to be disenchantment with having joint venture with government-owned enterprises. In forming joint ventures, the preference is more towards private sector. In a developing country like Myanmar, joint ventures may be preferable to foreign subsidiaries, particularly because the spillover effects are more easily reaped in the former as compared to the latter. Hence it is important to facilitate the emergence of a vibrant domestic private sector such that more foreign investment flows into the country.

Trade policy reforms and performance

Given the strong link between trade and investment which, as we have argued in Chapter 1, is apparently much stronger in ICT production as compared to most other industries, let us examine the trade policy framework in the country to explore the possibilities of further policy reform options, if any, to promote the production and use of ICT in Myanmar.

Trade policy: The present scene

Ever since the initiation of economic reforms in 1988, Myanmar has undertaken a series of measures to promote trade. These included, among others, permitting private enterprises to carry out exports and imports, promoting border trade, streamlining export and import procedures by lowering tariff and non-tariff barriers and reactivating the Myanmar Chamber of Commerce and Industry. As a founder member of WTO, its foreign trade policies are today guided by the rule-based multilateral trading system. In general, the registered importers/exporters have the right to export all the commodities except rice and rice products, which are reserved to be solely exported by the state-owned Economic Enterprises. The registered exporters are also allowed to enjoy 100 per cent export retention. Import is generally allowed against the export earnings with a view to promote exports and to overcome the trade deficits.

Since March 1998 the government imposed a restrictive import policy, requiring that all imported items must fall within either the priority A[15] or priority B lists. Priority A list items should be imported in a ratio of at least 80 per cent, and priority B list items can be brought in as a maximum of 20 per cent. B list items may only be imported after the arrival of the A list items. As of December 1998, the Ministry of Commerce stipulated that items which are not restricted but which do not fall within either the A or the B list also will be treated as list goods. Import permits may be obtained by producing evidence of export earnings. Another initiative taken by the government in July 2000 to contain the balance of payment situation involved imposing a remittance cap on earnings to $10,000 per month. Government also imposed a 10 per cent tax on all the exports. While discussing with Business people in Myanmar it was told that "we export to import". Such a practice is bound to affect the investment prospects and long-term growth of the economy. For example, a new venture, which involves import of machinery, might find it difficult to import because the export earning is yet to flow in, as project has not been commenced. Hence, it is worth considering doing away with such policy measures, which might be instrumental in achieving short-term goals at the cost of long-term growth.

Myanmar follows the Harmonized System of International Nomenclature. Three types of taxes can be levied on imports: import duties, commercial taxes and license fees. After joining ASEAN in 1997, measures have been undertaken to comply with the ASEAN Common Effective Preferential Tariff Scheme (CEPT). It is understood that Myanmar is in the process of meeting the CEPT tariff reduction commitments to be phased in between 2001 and 2008. In 2003, tariffs ranged from a low of zero to a maximum of 40 per cent, with cars, luxury items, jewelry and items produced in the country facing the highest tariffs. IT goods in general attract tariff less than 5 per cent. Most of the industrial inputs, machinery and spare parts, attract an average tariff of about 15 per cent.

The major impediment to trade, perhaps, relates to the foreign exchange regime in the country. Myanmar has a highly overvalued foreign exchange wherein the official exchange rate is about K 6 to a dollar, where as the market exchange rate is as high as 900 to a dollar. Such overvalued exchange rate coupled with high rates of inflation in the domestic economy has the effect of very high real effective exchange rates that undermine the export competitiveness. Foreign firms are also required to record transactions at the official rate when submitting forms to the government. When foreign firms bring in foreign exchange to be used for purchases from the local economy, they must deposit it in a state bank and withdraw any funds used in Foreign Exchange Certificates (FECs). Foreign firms sometimes avoid the official exchange rate by paying for services in dollars. While one needs to appreciate the economic imperatives that have induced the government to adopt the present foreign exchange policy, to the extent that the present exchange regime creates impediments to trade, it is advisable to explore the possibilities of reforming the exchange rate mechanism.

Trends in the external sector

There is a two-way link between trade policy and trade performance. While the trade policy has a significant bearing on the overall performance of the external sector of an economy, it is also likely that the developments in the external sector influence the policy changes to a great extent. Hence to appreciate the trade policy changes in Myanmar, it is appropriate to begin with an examination of the major trends in the external sector with focus on trade. Table 6.6 presents data on the trend in exports, imports and trade balance since 1980–81. Following observations could be made from the table. During 1980–81 to 1998–99, while the exports almost doubled, imports recorded more than threefold increase. The natural outcome has been the growing trade deficit recording more than sixfold increase during the period under consideration. Under such circumstances it is imperative for any economy to adopt certain measures to contain the widening trade gap. The problem could be addressed by either boosting exports or restricting imports. The approach adopted by Myanmar appears to be the latter one as is evident from the reduction in the annual rates of growth in imports recorded since 1998–99. Notwithstanding the growing trade gap, the overall balance of payment situation was kept under control through

Table 6.6 Trend in export and import in Myanmar

Year	Export	Annual growth	Import	Annual growth	Balance of trade
1980–81	3,225.1		4,635		−1,409.9
1985–86	2,653.9	−17.71	4,802	3.60	−2,148.1
1990–91	2,961.9	11.61	5,522.8	15.01	−2,560.9
1994–95	5,405.2	82.49	8,332.3	50.87	−2,927.1
1995–96	5,043.8	−6.69	10,301.6	23.63	−5,257.8
1996–97	5,487.7	8.80	11,778.8	14.34	−6,291.1
1997–98	6,446.8	17.48	14,366.1	21.97	−7,919.3
1998–99	6,755.8	4.79	16,871.7	17.44	−10,115.9
1999–2000	8,947.3	32.44	16,264.8	−3.60	−7,317.5
2000–2001	12,736	42.34	15,073.1	−7.33	−2,337.1
2001–2002	17,130.7	34.51	18,377.7	21.92	−1,247
2002–2003	15,289.6	−10.75	11,967.5	−34.88	3,322.1

Source: CSO, Different years.

borrowing, increased receipts on account of private services, remittances and FDI. Yet it has been argued that Myanmar had to manage very often with a foreign exchange cover just sufficient to meet one month's imports (ADB 2002). It is also important to note that during the terminal year (2002–2003) Myanmar recorded a trade surplus in balance of trade for the first time since 1980–81. While this might *prima facie* appear as a major achievement, import compression is likely to have an adverse effect on the future growth prospects of the economy to the extent that the growth depends on imported capital goods, raw material, spares and so on.

IV. Production and use of IT

Till recently, Myanmar Posts and Telecommunications (MPT), which employs about 20,800 persons, has been the monopoly provider of postal and telecommunications services in the country. Thanks to the initiatives by the government, the growth of telephone lines in the country has shown a marked increase during the post-1990 period. As compared to the recorded ACGR of 5 per cent during 1980–81 to 1985–86 and 7.8 per cent during 1985–86 to 1990–91, the growth rate during the third period (1990–91 to 1995–96) was almost twice (14.4 per cent) that of the previous period. Growth rate during the fourth period was also much higher (10.55) than that of the first two periods under consideration (Table 6.7). Also during 1980–81 to 2000–2001 the number of

Table 6.7   Growth of telecommunications in Myanmar

Year	Telephone Lines (number)	Exchanges (number)	Lines in Yangon
1980–81	46,374	206	28,625 (61.7)
1985–86	59,343	228	30,796 (51.9)
1990–91	86,333	317	45,575 (52.8)
1995–96	169,530	469	83,234 (49.1)
1996–97	199,017	493	98,736 (49.6)
1997–98	225,315	517	114,909 (51.0)
1998–99	240,673	528	123,674 (51.4)
1999–2000	260,579	539	136,676 (52.4)
2000–01	282,853	556	148,577 (52.5)

Source: CSO (2001).
Note: Figures in the parenthesis show share of Yangon in total.

telephone exchanges increased from 206 to 556. Table 6.7 also reveals that in 1980–81 almost 61 per cent of the telephone lines were concentrated in Yangon, but by 1995–96 the share of Yangon declined to 49 per cent, though by 2000–2001 its share showed a marginal increase to reach 52.5 per cent.

At the same time, going by the World Bank (2002) data, there is a waiting list of over 93,000 in the year 2000 and the average waiting time is as high as 5.3 years. The cost of obtaining a fixed line from MPT, K 500,000 in 2002, is beyond reach of majority of the population. Also, we cannot ignore the fact that the intra-national telecom divide yet remains as an issue to be addressed. Data presented in Table 6.8 reveals while the telecom density (fixed) is as high as 2.54 in Yangon, which in most of the states/divisions is much lower. To illustrate, Ayeyarwaddy, which has a population of 7.1 million, has only 14,970 telephones giving rise to a telephone density of 0.21. Rakhine recorded the lowest telephone density (0.17). Needless to add, addressing the issue of intra-national digital divide is expected to take the country miles ahead in its overall development process.

Mobile telephone

Myanmar is one of the pioneering countries in South Asia to provide cellular services which had its beginning as early as in 1992. However, for obvious reasons, the subscriber base grew at a rather slow pace – during the first four years the total number of cellular subscribers was only around 2000. Initially, Myama Post and Telecom was the sole provider

Table 6.8 Inter-regional variation in the teledensity in Myanmar

Regions	Telephones (number)	Population (million)	Density
Kachin	5,653	1.31	0.43
Kayah	1,374	0.30	0.46
Kayin	2,728	1.52	0.18
Chin	1,330	0.48	0.28
Sagaing	11,429	5.52	0.21
Tanintharyi	3,886	1.42	0.27
Bago (east)	9,236	2.90	0.32
Bago West	5,957	2.30	0.26
Magaway	12,637	4.87	0.26
Mandalay	32,623	7.14	0.46
Mon	7,274	2.55	0.29
Rakhine	4,749	2.81	0.17
Yangon	149,619	5.90	2.54
Shan (south)	6,961	2.11	0.33
Shan (north)	7,888	2.00	0.39
Shan (East)	2,894	0.77	0.38
Ayeyarwaddy	14,970	7.10	0.21
Total	281,226	51.00	0.55

Source: Kyaw Aye (2002).

of mobile services. The cost of mobile telephone charges by MPT is given in Table 6.9. It is obvious that in a country with per capita income of about $300, such high cost/prices are bound to stand in the way of diffusion of mobile technology. Hence there is the need for bringing down cost/price of mobile phone and the ways and means of achieving this form, the priority agenda for the policy makers.[16]

Given the fact that in a vast country with widespread settlement patterns and unfriendly geographical terrains, it is only natural that a single agency might find it difficult to meet all the telecom needs of the entire country. Hence, in 1998, a Service Provider Agreement was signed between MPT and the Iridium, Southeast Asia for providing satellite telephones. The operation of Iridium products and services in Myanmar, strictly in line with the guidelines laid down by the government, was started in 1999. Iridium used to charge US$3.9 per minute for both incoming and outgoing international calls with a weekly rental rate of US$70 and monthly rental rate of US$ 220.[17] Given the high cost of satellite telephone its viability has been questioned even in developed countries, and its contribution in Myanmar has been limited.

Table 6.9   Cost of mobile telephones in Myanmar (1999)

Item	Retail price US$
Handset	
Motorola ISU	5,200
Motorola Pager	750
Airtime	
Domestic	5.00
Regional	6.00
International from Myanmar	8.00
International outside Myanmar	5.50
Services voice	
Monthly fee	50.00
Activation fee	40.00
Reactivation fee	20.00
Deposit	750.00
Paging	
Monthly fee	100.00
Activation fee	40.00
Reactivation fee	20.00

Source: Lwin (1999).

Computers and Internet

The net result of the series of restrictions and high cost is that the use of Internet is restricted to a handful of privileged enterprises (Kyaw 2002). It was discerned during discussion with the senior officers that about 30 government ministries/departments are currently having PCs and most of them are networked and have Internet access. According to World Bank (2002) there are only about 10,000 Internet users in the year 2000, illustrating the fact that the benefits of new technology are restricted only to a privileged few.

On the whole, it appears that the series of efforts made by the government is not bringing the desired returns primarily because of the lack of complementary innovations. If the government were to achieve the declared objective of harnessing ICT for the overall development of the economy and society, there seems to be no other way other than relaxing the restrictions and following a more-open-door policy such that the present monopoly/duopoly system is replaced by a more competitive one. If the experience of India is any indication, opening up the telecom sector for the private sector coupled with imaginative institutional innovations for regulation has had the effect of providing rich returns in terms of drastic reduction in cost and significant increase in access to wider section of economy and society.

Present state of IT production

As already noted, the economic reforms initiated in the late 1980s underlined the need for promoting the private sector, especially in the manufacturing sector. While the share of manufacturing in GDP is only of the order of 10 per cent, the private sector contributes about 71 per cent of it. If we take productive sector of the economy as a whole, the share of private sector is about 87.2 per cent. Although the contribution of the private sector to the total GDP is significant, the number of large private factories/establishments is only about 142: the share of large factories/establishments among the total private factories/establishments is only 0.3 per cent (Table 6.10). Most of the private factories/establishments (95.4 per cent) are very small scale, that is below ten workers. This economy-wide trend appears to be holding good in the case of IT production as well.

During discussion with the representatives of the Myanmar Chamber of Commerce and Industry it was discerned that the IT goods production in the country was at a low level. While there are two local producers of computers, almost 70–80 per cent is accounted by the so-called "Grey market". While looking at the major areas of operations of the members of Myanmar Computer Industry Association, it was discerned that hardly any one is engaged in computer production. Most of them are engaged in hardware/software supply and service, and software production. In consumer electronics, MNCs like Toshiba and Daewoo are having operations in the country. In the public sector, Myanmar Machine Tool and Electrical Industries (MTEI) has one electrical and electronics factory

Table 6.10 Number of factories across different size and ownership categories 1998–99

Category	State-owned	Cooperatives	Private	Total
Below 10 workers	719 (45.6)	443 (65.4)	50,844 (95.4)	52,006 (93.7)
10–50 workers	291 (18.5)	175 (25.8)	2,134 (4.0)	2,600 (4.7)
51–100 workers	257 (16.3)	57 (8.4)	150 (0.3)	464 (0.8)
Over 100 workers	309 (19.6)	2 (0.3)	142 (0.3)	453 (0.8)
Total	1,576 (100)	677 (100)	53,270 (100)	5,523 (100)

Source: Government of Myanmar, Review of Financial, Economic and Social Condition, 1998–99.

located at South Dagon, with following product lines – Fluorescent Lamps and incandescent bulbs, electric rice cookers, electric irons, electric hot plates and dry cell.

While the production base in the IT goods appears to be limited, Myanmar has already initiated some bold steps towards creating software/service production in the country. This is manifested in the setting up of a STP at the instance of Myanmar ICT Development Corporation (a consortium of 50 private companies) with the active support and cooperation from the Government of Myanmar wherein the needed communication infrastructure is being provided by the publicly owned Bagan Cybertech. The project, set up in the Hline University Campus with a total investment of about K 2.5 billion, was initiated in March 2001 and the first phase was completed within a very short span of about 10 months and the park was inaugurated in January 2002. The first phase of the project has a developed area of about 11 acres with 32 rooms (100 ft × 50 ft). As of June 2003, the occupancy rate has been 100 per cent. The park has been able to attract two foreign companies; it is also the home for an e-learning center, an incubation center for promising local software programers and the Japan–Myanmar e-learning center. The activities of the units in the Park include software development, human resource development, national level projects, data processing services and consultancy services, and it provides employment to about 700 people. On the whole, the technology park experiment is a testimony to the positive outcome of public–private participation. It is heartening to see that the government is in the process of replicating this success by setting up another park at Mandalay.[18]

V. Concluding observations

Myanmar presents a typical case wherein the concerted efforts at developing an ICT production base and promoting its use notwithstanding, the returns remain limited mainly on account of the absence of complimentary innovations. The study finds that though the computer science law and the Master Plan have certain laudable objectives, a series of restrictions on the use of ICT have had the effect of mitigating the positive effect of various initiatives that the government has undertaken. While huge investments were made towards increasing the transmission coverage of TV and radios, low affordability on account of low-income levels coupled with restrictions in the use of TV and radio in the form of licensing seem to have an adverse effect. Similarly, while efforts were made towards modernizing the telecom infrastructure, lack of competition and resultant high prices/costs seem to have had led to the reduced access. Here the study point towards the need for creating a more competitive environment in the telecom sector and doing away with the licensing system for TVs and radios and restrictions on the use of Internet.

The trade policy regime has been characterized by a series of changes mainly aimed at restricting imports in the context of widening trade gap. While such reforms have had the effect of improvements in the current account, the study notes that import compression is likely to have adverse effect on the future growth prospects of the economy in general and the growth of ICT sector in particular. A major issue in the trade policy front related to the overvalued exchange rate regime. Hence the need for using exchange rate mechanism as an effective tool for managing the balance of payment. This, however, calls for, among other things, reforms in the financial sector and more specifically a change in the status of the Central Bank of the country.

Various improvements made in the foreign investment law notwithstanding, the present study underscores the need for more transparency with respect to the provision of tax exemptions beyond the initial three years. Also, it may be worthwhile to reconsider the present practice of providing tax exemption for profits on exports for all commodities. Given the fact that the barriers to enter the international markets vary across different sectors, the incentives offered need to have some relationship with the height of entry barriers faced by the exporters. An analysis of the trend in investment approvals and actual inflows shows that since 1996–97 there has been a steady decline in both approvals and actual inflows. A real reversal in this trend is yet to take place. This Study also shows that FDI accounts for bulk of the investment in almost all the sectors of the economy and the most preferred mode of FDI is the fully owned foreign subsidiary followed by joint venture with local private sector. But given the low financial and technical capability of the domestic private sector at present, it is unlikely that the large MNCs find it advantageous to join hands with the relatively weak private sector in Myanmar. The observed trends tend to suggest that in a context wherein lower labour costs and subsidies are taken for granted by the MNCs, countries like Mynamar need to focus on building up the complementary capabilities that the foreign firms look forward to. This calls for earnest efforts towards building up of an innovation system, which is at present in its infancy.

While the present production base for IT goods appears to be limited, concerted efforts have been made towards laying the foundation of a software sector. This has been accomplished by the setting up of a software park and another one is being planned to be set up at Mandalay. The series of initiatives made by the government towards IT education and building up IT manpower base is likely to provide rich dividends in the near future, provided a more conducive environment for the production and use of IT goods and services is created.

High entry barriers notwithstanding, it has been argued that there are real opportunities for countries like Myanmar to enter into those areas of ICT goods like passive and electro-mechanical components and entertainment equipment like TVs, radio and two-in-ones which are characterized by relatively stable technology and low investment requirement, provided restrictions on trade and investments are removed. In the sphere of ICT services, in the short run it may be rewarding to develop less skill-intensive ICT-enabled services like data entry, medical transcription and call centers so as to create new employment opportunities. At the same time, there is the need for conscious efforts towards skill empowerment such that the economy does not get locked up in low/medium technology activities, and an upward movement along the skill spectrum is ensured.

7
Vietnam: Another Tiger in the Making?

I. Introduction

Ever since the initiation of the policy of renovation (Doi Moi) in 1986 aimed at transforming an economy, which has been described as "thirty years behind Taiwan, twenty five years behind Malaysia, twenty years behind Thailand, twelve years behind China and eight years behind Indonesia in terms of overall development" (Chu Huu Quy 1996: p. 18) by creating a market-oriented economic environment, the economy of Vietnam witnessed major changes in terms of growth and structural change.[1] During 1986–90, GDP recorded an annual average growth rate of around 4 per cent followed by a much higher growth rate of 8.9 per cent during 1991–96. While the Asian financial crisis humbled even the star performers of Southeast Asia, the economy of Vietnam managed to traverse through the troubled waters recording an ACGR of 6.5 per cent during 1996–2001.[2] The tempo of growth has been maintained in 2002 and 2003 with a recorded annual growth rate of 6.4 and 7.1 per cent respectively (ADB 2004). To go beyond the growth rates, rate of inflation, though accelerated, stayed below 10 per cent. The share of exports and imports in GDP increased from 22 per cent in 1990 to 50 per cent and 62 per cent respectively in 2001 and 2003 and the level of depreciation of Dong was much low as compared to other Asian countries. Population under the poverty line declined from about 58 per cent in 1993 to 37 per cent in 1998.

Commentators argued that Vietnam has been mildly affected by the Asian crisis more to do with controls and regulations than with sound economic fundamentals. Hence it was predicted that Vietnam economy would gradually slide into a deeper recession (Kokko 1998).

If the available data for the later years is any indication, such predictions have turned out to be fallacious as there has been a turnaround in balance of trade, decline in the rate of unemployment to a level as low as 1.7 per cent in 2003 (ADB 2004), improvement in the inflows of FDI and marked revival in the growth rates since 1999. While the growth dynamism has been coincided with structural changes in the economy, the share of state sector, which is valuable as a provider of employment and key player in key sectors of the economy, yet vulnerable on account of its poor financial performance,[3] has shown only a marginal decline from about a little over 40 per cent in 1995 to 38.5 per cent in 2001.

Among the policy reforms addressing the domestic economy, an important one related to the emphasis on IT. The Party Politburo concerning science and technology, as early as in 1991, underlined the need to concentrate efforts to develop some spearhead scientific and technological areas such as electronics and informatics. Since then there has been a series of initiatives by the government towards building an innovation system with a view to promoting the production and use of IT in the country. Simultaneously, trade and investment regime in the country has also been subjected to unprecedented reforms. As a result, Vietnam today is integrated with the other regional economies and the global economy more than ever before. This is manifested in Vietnam becoming a full member of ASEAN in 1995 and its ongoing efforts to get entry into WTO. In 2001 the government signed the US–Vietnam Bilateral Trade Agreement (BTA)[4] with a view to have normal trade relations with the major player in the global economy. With the BTA in force, US tariffs on imports from Vietnam have dropped from an average of 40 per cent to an average rate of 3 per cent, allowing Vietnamese exporters to compete on equal or nearly equal terms in the US market.[5] In return, Vietnam has undertaken to open its market and reform its legal system in a variety of areas, including telecommunications and the IT sector.

II. NIS governing the production and use of IT

Realizing the importance of science and technology in transforming the economy of Vietnam, in 1997 the Ministry of Science, Technology and Environment approached the IDRC and the Canadian International Development Agency (CIDA) for assistance in developing a new science and technology strategy. Accordingly, a team of experts appointed by IDRC and CIDA undertook a detailed analysis of the S&T system in Vietnam and came up with a series of recommendations pertaining to institutional interventions and policy measures to revamp the NSI of Vietnam.[6] Reforms undertaken since then by Vietnam have been broadly in tune with these recommendations made by the team of experts.

Bulk of S&T activities in Vietnam are undertaken by the national centers for R&D and the agencies attached to different ministries, universities and other institutions of tertiary education with limited contribution by R&D scientists and engineers working in industrial enterprises, though there have been some changes in the recent years. The general institutional setup for R&D as identified by Bezanson *et al.* (1999) may be described in terms of three main components:

1. *Laboratories and other R&D units within the government ministries or under the control of government agencies*– About 180 such R&D units existed in 1999, located in various parts of the country, although most of them were in the two metropolitan areas.
2. *Universities and other higher-education departments that perform research as part of their normal activities* – Only a limited number of faculties and academic departments at Vietnam's universities and colleges were found having the personnel, equipment, libraries and other resources needed to undertake serious R&D. Among these, the campuses of the national university and the two largest polytechnic universities are the most research intensive. Thus a research-based university system in Vietnam has been emerging only slowly.
3. *The national institutions for research that are not directly under an individual government ministry or agency*– These are designed to act as national networks of S&T and are placed under the Government Office (that is, the Office of the prime minister). The most significant of these national institutions is the National Center for Natural Science and Technology (NCNST), with northern and southern branches and facilities in some other parts of the country. Originally modeled after an academy of sciences, it was restructured in 1993 to become more like a center for applied research and experimental development. It performs advanced basic research mainly in two areas: mathematics and theoretical physics.

These three main components of Vietnam's national R&D structure are expected to have close links with each other. The functional differentiation allocates applied research and experimental development to the laboratories of individual ministries, whereas the universities and colleges are the prime producers of highly specialized human resources for R&D, and the NCNST has the prime responsibility for the most advanced forms of research and for R&D if no specific ministry is a customer (Bezanson *et al.* 1999).

Earlier Vietnam's approach has been one wherein specific technological solutions have been proposed by the government, in accordance with the expertise and resources available, but the government avoided setting specific priorities within whole fields of technology. But by the mid-1990s, a clear ranking emerged in the form of four national priority programs for high technologies like IT, biotechnology, new materials and automation of which the most comprehensive program has been for IT. High-level inter-ministerial committees were appointed to monitor and coordinate the national programs in these high-tech areas.

Policy initiatives and institutional interventions in ICT

Ever since the explicit recognition by the Politburo on the importance of electronics and informatics in transforming the economy and society of Vietnam, there has been a series of policy initiatives and institutional interventions towards promoting the production and use of IT. Table 7.1 presents a select listing of such initiatives. It is evident from the table that there is hardly any field of IT which has failed to attract the attention of the government and in wanting for initiatives. What follows is an examination of the select initiatives having bearing on the production and use of IT in Vietnam at present. Such a selective approach has induced us to focus on the Directive No. 58-CT/TW, the Action Plan that followed and a few select decisions having direct

Table 7.1 Policy initiatives for the production and use of IT in Vietnam

Year	Decisions/decrees
1991	Resolution No. 26-NQ/TW by the Politburo called for developing some spearhead scientific and technological areas like electronics and informatics
1993	Government resolution No. 49/CP concerning the development of IT in Vietnam in 1990s
November 29, 1996	Decree No. 76/CP of the government on guiding the implementation of a number of provisions on copyrights in the civil code
June 5, 2000	Resolution of the Government No. 07/2000/NQ-CP of establishment and development of the Software industry in the 2000–2005 period

October 17, 2000	Directive No. 58-CT/TW on accelerating the use and development of information technologies for the cause of industrialization and modernization
May 21, 2001	Circular No.31/2001/TT-BTC of the Ministry of Finance on the instructions of the implementation of preferential tax treatment regulated at Decision No. 128/2000/QD-TTG of the prime minister
May 24, 2001	Decision No. 81/2001/QD-TTG of the prime minister on approving the Action Plan to implement the Directive NO. 58-CT/TW of the communist party of Vietnam
August 23, 2001	Decree No. 55/2001/ND-CP of the government on the management, provision and use of the Internet services
November 9, 2001	Decision on the establishment of the Steering Committee for Action Plan to implement the Politburo Directive No. 58-CT/TW
February 8, 2002	Decision No. 33/2002/QD-TTG of the prime minister on the approval of Vietnam Internet Development Plan in the 2001–2005 period
February 20, 2001	Decision No. 19/2001/QD-TTG of the prime minister on adding computer products to the list of key industrial products
March 21, 2002	Decision No. 44/2002/QD-TTG of the prime minister on using the electronic documents as accounting documents for calculation and payment capital in payment service supplying organizations
May 29, 2002	Decision No.543/2002/QD-NHNN of the Governor of the State Bank of Vietnam on issuing, controlling and using digital signatures on electronic documents in inter-ban electronic payments
July 25, 2002	Decision No. 112/2001/QD-TTG of the prime minister on ratifying the project on the state administrative management of Computerization in the 2001–2005 period
October 18, 2002	Decision No. 158/2001/QD-TTG of the prime minister on the approval of Vietnam's post and telecommunication development strategy until 2010 and orientation until 2020
November 20, 2002	Decision No. 128/2000/QD-TTG of the prime minister on a number of policies and measures to stimulate investment and development of software industry

Source: Compiled by the author from different official sources.

bearing on software development, investment promotion and creation of IT infrastructure.

Directive No. 58-CT/TW [7]

This directive issued by the Central Executive Committee of the Communist Party of Vietnam in 2000 forms the cornerstone of all the subsequent initiatives made by the government. The document begins with an evaluation of the various initiatives made by the government since 1991 for promoting the production and use of IT in the country. Highly realistic document does not mince words when it self-critically sums up the status of IT use in 2000 in the following words:

> The status of IT use in Vietnam, however, is still backward. Progress has been slow, creating the risk of a growing gap compared to many countries in the world and in the region. IT use and development has not met the requirements of Industrialization and modernization and the needs for international and regional integration. Human resources in IT have not been substantively prepared in terms of timeliness, quantity, quality and ability in foreign languages. Telecommunication and the Internet are not convenient for users, and they have not met the requirements of IT use and development in respect of speed, quality and cost. Investment for IT has not been sufficient. The state management of IT is still scattered and inefficient. The IT use of some places is still wastefully formal and unrealistic. (p. 3)

The Politburo recognized the ubiquitous nature of IT and the imperative of harnessing IT for the overall transformation of economy and society. Hence it called for increased IT use in every sector, building up of national information network and making IT industry a spearhead economic sector with a growth rate higher than any other sectors of the economy. The Politburo also requested all the sectors to broadly and effectively use IT in the whole of society, accelerate the training and utilization of human resources for IT use and development and renovate the state management of IT sector.

A careful reading of the document will reveal that there is hardly any area relevant to IT production and use that remain untouched by it. The directive, however, may be considered a vision document and to understand the extent to which the vision is being translated into reality, one needs to have an examination of the Action Plan prepared by the government.

IT Action Plan 2000–2005 [8]

Reiterating the vision highlighted in the Directive No. 58, the Action Plan (2001) presented the goals and targets to be achieved in the medium term (by 2005) in more concrete terms as follows:

- The level of IT application for the country as a whole by 2005 shall be of medium level when compared to that of other countries in the region. But the same with respect to the activities of the party and state management agencies and main spearhead economic branches of the national economy in the central government shall be comparable to the advanced countries in the region.
- To develop Internet and telecommunication networks with modern technologies, higher bandwidth, speed and quality to provide the society and consumers with a variety of services with same or even lower prices as compared to the average in the region. By 2005, the provinces and cities all over the country will be linked by fiber cables and the rate of Internet users will be of 4 or 5 per cent of the total population.
- IT industry will achieve the annual average growth rate of 20–25 per cent, supporting key industries to develop and ensuring the high and sustainable economic growth rate.
- 50,000 IT experts at different levels will be trained of which 25,000 will be high-ranking experts and professional programers with good command of foreign languages.

Keeping the above objectives, the Master Plan called for:

Promoting the IT use in prioritized sectors: The sectors include, but not limited to, agriculture and industry, national security and defense, state administration, the party and state agencies.

Developing the national network of telecommunication and Internet: This aimed at providing greater access and high quality service and opening the Internet market to allow many Internet Connection Providers (IXPs), ISPs and Operational Service Providers (OSPs), leading to greater competition and lower cost. It was targeted that by 2005 the number of Internet users should increase tenfold as compared to the 2000 level resulting in an Internet penetration rate of 1.3–1.5 subscribers per 100 people and an Internet user rate of 4–5 per cent. It was also envisaged that from 2002 to 2003, all of institutes, universities, colleges and vocational high schools will be able to have Internet access. It was aimed that by

2005, 50 per cent of senior secondary schools, all the hospitals under the central government and more than 50 per cent of the provincial hospitals will have access to Internet. Further, by 2005 all the ministries, branches and administrative agencies of the central state management mechanism and provincial and district government will be linked to the government's Wide Area Network (WAN) and Internet.

Developing the IT industry: The Master Plan envisaged a total turnover of $500 million in software of which exports were to account for about $200 million. To achieve this target the government planned to invest $50–70 million by way of human resource development, marketing and product development projects. The Master Plan also envisaged the production of computer and communication equipment at least to the extent of 80 per cent of domestic demand.

Development of human resources: The Action Plan, while aiming at improving the quality and quantity of existing education and training system, also had provisions to send annually about 300 students and 500 officers at all levels abroad for short-term training in IT. Also it was envisaged that about 20 per cent of the government officials will be trained and provided with more IT knowledge.

Accelerating R&D activities in IT: This involved setting up of commercially viable technological incubators with a view to catching up with technological achievements in the world and implementing efficiently the adaptation and transfer into Vietnam.

Creating favorable legal environment for IT use and development: This aimed at, among others, protection of intellectual properties and copyrights, maintaining IT security and privacy.

Perfecting state management system in IT area: This involved organization of a unified state management system in IT and telecommunications with necessary mechanisms, including a professional title system for information/IT managers (CIO).

Raising the awareness of IT in the society, especially, for the state leaders and managers: This involved, popularization of the knowledge of IT and the information society through televisions and other mass media and short-term training on IT for leaders of all levels. The Action Plan, along with setting the target, also laid down certain key programs and projects, which are exhaustive by any standards, and clearly assigned the responsibility of implementation to different agencies concerned as specified in Table 7.2.

Table 7.2 Key programs and projects identified by the Action Plan in Vietnam

Key programs/projects	Implementing agency
Developing and improving telecommunication and Internet infrastructure	Department General of Post and Telecommunications
Developing IT human resources	Ministry of Education and Training
Establishing and developing the software industry	Ministry of Science Technology and Environment
Establishing and developing hardware industry	Ministry of Industry
Improving and upgrading party's information system	Central Office of the Communist Party of Vietnam
Modernizing the banking system	State Bank of Vietnam
Completing the Financial Information System	Ministry of Finance
Modernizing the Customs Information System	Department General of Customs
Improving the state's Statistical Information System	Department General of Statistics
Organizing and developing e-commerce	Ministry of Trade
Using IT for industrializing and Modernizing Agriculture and Rural Development	Ministry of Agriculture and Rural Development
Developing a number of Pilot Information Systems for Urban management in Hanoi and Ho Chi Minh City	People's Committee in respective cities
Using and developing IT in National Defense	Ministry of Defense
Using and developing IT in National Security	Ministry of Police
The Electronic Information System on science technology, education-training and culture society	Ministry of Culture and Information
Improving the Electronic Information System on law	Ministry of Justice

Source: Decision No. 81/2001/QD-TTG, dated May 24, 2001, of the prime minister on approving the Action Plan to implement the Directive No. 58-CT/TW of the Communist Party of Vietnam.

Resolution on the establishment and development of software industry

The resolution is based on the premise that Vietnam has several favorable basic conditions for the development of software industry like an expanding IT world market, minimal initial investment requirements, quick application of IT by the Vietnamese users and many experienced IT specialists of Vietnamese communities in foreign countries who want to cooperate and invest in the IT field in Vietnam. The resolution states that initially the attention shall be given to exports through contract-manufacturing and providing service. At the same time, the resolution calls for expanding the domestic market and an approach of import substitution with respect to software in some critical sectors. Target for software production has been set at $500 million by the year 2005 and with a view to achieve these targets, the resolution called for offering massive incentives for the software industry which included additional incentives for FDI in software[9] exemption from value-added tax, zero per cent export tax, preferential rates of corporate income tax, the provision to refund a part of the corporate income tax for re-investment and preferential treatment for those earning higher levels of income from software industry.[10] Also to tap the potential of overseas Vietnamese it was envisaged that the overseas Vietnamese coordinate with relevant ministries and branches to create favorable conditions such that they could cooperate with and assist domestic organizations and individuals in market development and technology transfer.

In tune with recommendations of the IDRC-CIDA expert team STPs have been set up in different parts of the country with a view to achieve the targets for production and export of software and services. The major parks in operation are

- Saigon Software Park
- Danang Software Park
- Quang Trung Software Park
- Hai Phong Sofware Park
- Hue Software Park
- Can Tho Sofware Park and
- Hoa Lac Hightech Park

By 2003, Quang Trung and Saigon Software parks were running at full capacity, and are full-service software parks. The Quang Trung STP had raised nearly $4 million in revenue and is home to some 51 firms employing about 1000 software engineers. The Saigon STP has about 16 companies and more than 600 staff. Some other parks like Can Tho

STP caters primarily to training needs of provincial officials and remains more of a training unit. Some others like Danang STP (DSP), having both advanced training facilities and software development facilities, boast of strong in-house software development units and offer a host of services in training, software product development, network management content management and technology consultancy.

Human resource development for IT

The importance of human resource development repeatedly appears in most of the policy resolutions pertaining to IT, indicating the high priority rightly assigned to it by the government. As already noted, the government has a target of training over 50,000 IT specialists at different levels, of whom 25,000 are high-level programers fluent in English. For IT training there are 20 IT faculties in Vietnam's various universities and colleges, 45 technical colleges with IT programs and about 67 vocational schools with IT subjects. It has been estimated that in 2000 there were approximately 20,000 IT professionals (with Bachelors' degrees) in Vietnam, with 10,000 working directly in the IT industry on R&D or in education services (USAID 2001).

There are a number of bilateral agreements to promote the IT training in Vietnam. A joint working group on IT and electronics between India and Vietnam has been in existence since 1999. The Prime Minister of India during his visit to Vietnam in January 2001 announced a grant of Rs 100 million for a software and IT training center in Vietnam. In November 2001, the Government of Vietnam allocated this grant to Hanoi People's Committee for utilization in a US$4.4 million, Hanoi IT Transaction Centre project.[11] In November 2001, a Vietnam–Japan e-learning Center was opened in Ho Chi Minh City (HCMC) as part of cooperation between the two countries in IT. Under an MOU signed between the two countries in August 2001, Japan's Ministry of Economy, Trade and Industry had agreed to provide US$150,000 a non-refundable aid to Vietnam for IT courses. The Republic of Korea plans to send about 50 government officials to Vietnam to provide training in the latest IT techniques.

Government also has been successful in attracting investment in the field of IT training. A number of private schools have entered the Vietnamese market with the specific aim of developing IT professionals. The Royal Melbourne Institute of Technology (RMIT) has opened a school in HCM City, with plans to build a large campus near HCM City.[12] APTech, Tata Infotech and NIIT, the three leading private training institutes from India with operations in many countries, are also providing

IT-related programs in Vietnam. In addition to these formal educational institutions, there is considerable IT-related training provided by IT business associations. The Vietnam Association of Information Processing (VAIP) has been instrumental since the early 1990s in providing IT-related awareness over public TV and in setting up 90 IT training centers throughout Vietnam. These centers issue approximately 1000 certificates a month. In addition, the VAIP sponsors an annual IT Olympiad with participation from each university in an effort to promote student IT learning. The Vietnam Chamber of Commerce and Industry (VCCI), through its two chapters in Hanoi and HCM City, organizes wide array of IT-related workshops, seminar and training programs for its members and the business community. So far the VCCI is said to have offered 700 courses for SMEs, of which 70–80 were directly dealing with IT.

While all go well with quantity, what matters in a highly skill-intensive and competitive field like IT and software development is the quality of manpower. The study by USAID (2001) quoted the results of a study conducted by the Political and Economic Risk Consultancy Ltd, in which a Human Resource Index for Asian countries was developed. If the results of this study are any indication, all is not well with respect to the quality of IT manpower. The study developed an index on a ten-point scale variables such as overall quality of local education system, availability of high quality production labor, availability of high quality clerical staff and high quality management staff, English proficiency and high-tech proficiency. Vietnam was ranked low in virtually all categories (e.g. none were above 3.50 on a scale of 0–10), with the high-tech proficiency ranked 2.50 – the lowest of all countries included in the survey! English proficiency was also ranked the lowest of all countries (including China) with a ranking of 2.62 out of 10. A consolidated index of the education and human resource environment in Asia rated Vietnam at 3.79 out of 10, with Indonesia being the only country with a lower score than Vietnam (South Korea was rated at 6.91, Singapore 6.81, Japan 6.50, Taiwan 6.04, India 5.76, China 5.3, Malaysia 5.59, Hong Kong 5.28, Philippines 4.53, Thailand 4.04, Vietnam 3.79 and Indonesia 3.44).

In a country with limited experience in developing human capital for frontier technologies, it is not surprising to have such instances, at least in the short run. But this is not affordable in the long run because of the potential threat of undermining the development of IT sector on the one hand and "educated" unemployment on the other. Therefore, there is the need for targeted interventions to strengthen the innovation system by nurturing strong linkages between the academia and the industry, which is generally considered as win–win for both. Also it is worth considering the cooperation between countries like India to ensure the supply of high quality IT Manpower.

Information infrastructure

Radio and television

There are three major radio stations in the country and one national broadcaster called the Voice of Vietnam. The Voice of Vietnam, fully owned by the government, is the official network of the government and it broadcasts on AM, FM and SW. The national broadcasts are not only in Vietnamese but also in ethnic tongues like Khmer, H'mong Ede GiaLai and Bana. At present, Voice of Vietnam has provincial radio stations at 61 provinces primarily using AM while Hanoi and HCM City are having FM stations. According to the data furnished by the ITU and the World Bank the number of radios per 1000 people in 2001 was 109 and this has to be compared with 139 for the low-income countries, 112 for the South Asia and 235 for Thailand.

Similar to radio, the TV telecasting is also a state monopoly – Vietnam Television. The Vietnam Television has three nationwide channels as well as a local channel in each province. There is also the satellite channel mainly addressing the overseas Vietnamese. According to the data furnished by the ITU and the World Bank the number of TVs per 1000 people in 1997 was 47 and it was a little more than one-fifth (285) of Thailand. It was also found to be much lower than the recorded level for the low-income countries in general (138) and even the South Asian countries (61). However, the scene seems to have changed significantly within five years. By 2002 the number of TVs per 1000 people increased more than fourfold to reach a level of 197/1000. Thus with respect to the diffusion of TVs, Vietnam stood head and shoulders above the lower-income countries (91) or even the South Asian countries (84), and the gap between Thailand has declined significantly. According to ITU (2002d) there are approximately 10 million TV households meaning that around 80 per cent of Vietnamese homes have a television set. The TV penetration is much higher in the urban areas – around 96 per cent of urban households in Hanoi and 92 per cent in HCMC own a television set.

Telephones

Traditionally, the telecommunication service in Vietnam has been mainly under the control of the state-owned Vietnam Post and Telecommunications Corporation (VNPT). In June 2002, the then

General Department of Posts and Telecommunications (GDPT) permitted VNPT to separate its post and telecommunications divisions, ending the traditional cross-subsidization of postal services with telecommunications revenue,[13] and announced that the long-standing monopoly of the state-owned enterprises would slowly be phased out and competition introduced. On May 25, 2002, the Standing Committee of the National Assembly approved the introduction of Ordinance (43/2002/PL-UBTVQH10 on Post and Telecommunications) to replace the outdated Decree No. 109/1997/ND-CP on Posts and Telecommunications. This Ordinance provides a more transparent regulatory regime for private and public networks, dealing with issues such as interconnection, fees, charges and licensing. In July 2002, the National Assembly of the Socialist Republic of Vietnam decided to establish the Ministry of Post and Telematics (MPT). MPT is in the overall charge of policy making and regulatory matters on post, telecommunications technology, Internet, electronics industry, radio frequency management and the nation-wide information infrastructure. All these initiatives were aimed *inter alia* at achieving the official target of a teledensity between 8 and 10 telephones per 100 people by 2005, and between 15 and 18 by 2010.[14]

With a view to getting access to foreign technology and capital, VNPT also established a number of Business Cooperation Contracts (BCCs) with foreign telecommunication companies (Table 7.3). Though it is a fully state-owned enterprise, it operates like a holding company with many subsidiaries to carry out virtually every aspect of telecommunications (ITU 2002d). Vietnam also has a policy of allowing different government ministries to offer telecommunications services with a view to create a competitive environment. The Defence Ministry has already entered through Military Electronic Telecommunications Company (Vietel) and offers long-distance telephone service at much lower cost.

Internet and computers

The Internet service in Vietnam officially had its beginning in November 1997.[15] The use of Internet in the country is currently governed by the Decree No. 55/2001/ND-CP of the government on the management, provision and use of Internet services. The basic principles that guide the Internet development in Vietnam are

• The management of capacity must be in line with development requirements, while at the same time consistent measures must be taken to prevent abuse of the Internet to affect the national security and break national values and traditional good customs.

- The Internet should be developed fully with all high quality services provided at reasonable prices to meet with requirement of the national cause of industrialization and modernization.
- Scientific research, education, training, healthcare and software development institutions and organizations and part and state organizations will be prioritized in terms of the allocation of investment capital and application of financial support mechanisms regarding the provisions and use of Internet services.

In addition, the Decision of the prime minister (No. 158/2001) on approval of the Vietnam's Post and Telecommunication development strategy until 2010 and orientation until 2020 also contains a number of provisions regarding the growth and development of Internet. It has been envisaged that by 2010 at least 30 per cent of the subscribers will have access to broadband telecommunication and Internet. The strategy aims at mobilizing about US$11–12 billion during 2001–2020 for investment in the development of post, telecom and IT, of which about 50 per cent is expected to be mobilized during the first 10 years

Table 7.3 Business cooperation contracts for telecommunication services in vietnam

Company	Date	Number of lines	Value (US$ million)	Remarks
Japanese Consortium (NTT, Nissho Iwai, and Sumitomo)	November 1997	240,000	208 plus $14 Service contract	Northern part of Hanoi
France Telecom	November 1997	540,000	492.5	15 years contract, HCMC region
Cable and Wireless (US)	November 1997	250,000	207	
Telestra (Australia)	Several contracts beginning in 1988	NA	237 in 1988 subsequently increased	
Comvik (Sweden)	1993	NA	NA	GSM moblie

Source: ITU (2002d).

wherein the domestic resources accounting for about 6 per cent and the remainder from foreign investment.

III. Trade and investment regime

Investment policies

During the post-independence period, which marked the soviet style planning, foreign investment was not permitted in Vietnam. History, however, shows that during the period following Industrial Revolution, Vietnam has had substantial investment from France.[16] The revival of interest in FDI began with the initiation of Doi Moi (Renovation) policy with focus on market-oriented economic management. Almost a year after the initiation of the Doi Moi, in December 1987, the first foreign investment law was passed. This law has been revised twice and after almost 10 years, in November 1996, a new investment law was adopted which was considered as one of the most liberal in the region (Freeman 2002). The law specified different forms of foreign investment like 100 per cent foreign investment, joint ventures, Build–Operate–Transfer Contract[17], Build–Transfer–Operate Contract and[18] Build–Transfer Contract.[19]

The law was amended in June 2000 and a decree passed on July 2002 providing detailed regulations on the implementation of the law on foreign investment in Vietnam, which today governs the inflow of FDI into the country. Let us deal with it at some length.

The decree provides for all forms of FDI specified in the foreign investment law of 1996. In the case of joint ventures the legal capital must be not less than 30 per cent of the invested capital. But depending on technology, market, business results and other socio-economic benefits of the project, the investment license–issuing body may consider and permit the foreign joint venture party to have a lower capital contribution ratio but not less than 20 per cent of the legal capital. It appears that the current law makes cross-border mergers and acquisitions difficult if not impossible. Foreign investment directly into existing local companies, rather than as joint ventures or fully owned foreign subsidiaries, is currently constrained by the law. The local companies may currently issue shares to foreign investors only if they operate in one of 35 specified sectors. Even then, the approval of prime minister's office is needed before a foreign investor takes even minority equity (up to 30 per cent) stake in a local firm. This also potentially constrains the aggregate scale of FDI inflows and hinders local companies in tapping foreign capital and other non-financial inputs from overseas investors (Freeman 2002). If the evidence presented in UNCTAD (2000) is any indication, not only mergers and acquisitions dominate FDI flows in developed countries, they have also begun to take hold as a mode of entry into developing countries and economies in transition.

Article 80 deals, among others, with the transfer of technology. It states that the Government of the Socialist Republic of Vietnam shall create favorable conditions and shall protect the lawful rights and interests of a party transferring technology into Vietnam. Given the fact that there are different modes of technology transfer and the ultimate effect of technology transfer on the domestic technology capability building process of any economy depends to a great extent *inter alia* on the nature of technology transferred coupled with the terms and conditions attached, it is important that the law specifies in more detail the terms and conditions to be followed in transferring foreign technology. Also, as argued in Chapter 1 the process of technological change in developing countries is a combined outcome of technology transfer and in-house R&D. Hence it is important that technology transfer is complemented with in-house R&D activities. To promote R&D, the government might consider imposing a tax on payment for technology import and the resources thus mobilized being utilized for promoting domestic R&D activities.

Regarding recruitment of employees, it is stated that the foreign enterprises can directly recruit Vietnamese employees only if the Vietnamese Labor Supply Organization fails to recruit within 15 days. While the restriction to provide preference for Vietnamese in recruitment is understandable from the perspective of creating employment opportunities and skill formation for Vietnamese, restrictions on the employers' freedom in selecting the employees, even from among the Vietnamese, may have to be reconsidered.

As already noted, the Vietnam is perhaps the only country in the region having a specific government decision regarding the ways and means to attract FDI into the area of software wherein preferential treatment has been assigned in terms of lower tax levels, zero import duty on imports, zero export tax, preferential land use and rent, and also provision for preferential credit.[20] Moreover, Decision No. 19/2001/QD-TTG of the prime minister on adding computer products to the list of key industrial products also facilitates investment into the area of hardware production. On the whole, notwithstanding certain marginal issues, it goes without saying that the present laws governing FDI in the country in general and those guiding investment in the IT sector in particular provide the most conducive environment for attracting investment into the country.

FDI performance

The quality of FDI policy pursued by a country can be best gauged by an examination of the extent to which the country has been able to achieve the declared objective of the policy – attracting investment in the highly competitive world of FDI. We shall begin with an examination of the trends and patterns on FDI in the economy as a whole and finally present the situation in the IT sector.

Table 7.4 presents data on the trend in FDI approvals and actual inflow. Given the fact that the data provided by the Ministry on actual inflow is found to be at variance with that provided by UNCTAD, we have presented the data from both sources. We have also worked out the fructification rate (ratio of actual inflow to commitment) using both data. From the table it is evident that the FDI in Vietnam had its peak in 1996 (in terms of commitment) and in 1997 in terms of actual inflow regardless of actual inflow data provided by UNCTAD or the Ministry. In fact it has been found that in 1996 the FDI inflow as a per cent of Gross National Product (GNP) was the second highest in the world and

Table 7.4 Trend in FDI commitment and actual inflow in Vietnam (in US$ million)

Year	No. of projects	Commitment	Actual inflow		Fructification (%)	
			UNCTAD	Ministry	Ministry	UNCTAD
1988	37	371	NA			NA
1989	68	582	NA	100	17.17	NA
1990	108	839	180	120	14.30	21.45
1991	151	1,322	NA	375	28.36	NA
1992	197	2,165	NA	492	22.73	NA
1993	269	2,900	NA	931	32.10	NA
1994	343	3,765	NA	1,946	51.68	NA
1995	370	6,530	1,780	2,343	35.88	27.26
1996	325	8,497	1,803	2,518	29.63	21.22
1997	345	4,649	2,587	2,822	60.70	55.65
1998	275	3,897	1,700	2,214	56.81	43.62
1999	311	1,568	1,484	1,971	125.70	94.64
2000	371	2,021	1,289	2,043	101.07	63.77
2001	502	2,503	1,300	2,100	83.90	51.94
2002	802	1,621		1,250	NA	NA
2003	748	1,899		2,650	NA	NA

Sources: General Statistical Office (2001) and UNCTAD (different years).

the highest in the region (Freeman 2002).[21] Since 1996 there has been a decline in both commitments and disbursements; the commitments reached a low level of $1568 million in 1999 and there appears to be a revival since then to reach a level in 2003 comparable to that of 1996. It needs to be noted that such a revival is yet to take place in any of the other ASEAN New Comers.

The most distinguishing point is the extent of fructification. During 1995–2001 for which comparable data is available, the fructification rate is as high as 54 per cent according to the ministry data and over 40 per cent according to the UNCTAD data. The difference notwithstanding, the observed rate of fructification is high when compared to neighboring countries like Lao PDR (15 per cent), Myanmar (29 per cent) or large countries like India (27 per cent) and even comparable to that of China (50 per cent).[22]

FDI in IT and manufacturing sector

With a view to draw some inference on the importance of FDI across different manufacturing industries, we have estimated the change in the share of foreign enterprises in the output of manufacturing sector (at two-digit International Standard Industrial Classification (ISIC)) and estimated the ACGR in the output of local and foreign sectors. The results presented in Table 7.5 may be viewed with caution as the data presumably refers to the organized sector and may not represent the unorganized sector, which is known to be playing a key role in the production of IT goods like computers.

It may be noted that there has been a significant increase in the share of foreign sector in almost all the industries and among them the IT goods–producing sector recorded very high increase in their share. The share of foreign sector in the computing equipment industry increased from a little over 2 per cent in 1995 to over 96 per cent in 2001. In the case of TV, radio and communication equipment also the share almost doubled in a short span of six years from 46 to 81 per cent. The observed increase in the IT goods becomes spectacular when seen against the fact that for the manufacturing sector as a whole, the increase in the share of foreign firms was about 11 per cent (from 18 per cent in 1995 to 29 per cent in 2001). The ACGRs also tend to suggest that the foreign sector recorded growth rates much higher than the local firms in general and the IT sector in particular. A study by Binh and Houghton (2002) has shown that BTA will boost FDI into Vietnam by as much as 30 per cent initially.

Table 7.5 Relative performance of foreign and local firms in the manufacturing sector of Vietnam

ISIC	Industries	Share of foreign firms (%)		ACGR (1996–2001) (%)	
		1995	2001	Foreign	Local
15	Food products and beverages	19.10	20.56	11.90	10.24
16	Manufacture of tobacco products	0.14	0.57	37.88	8.83
17	Manufacture of textiles	17.33	28.25	20.46	8.44
18	Manufacture of wearing apparel	18.17	25.09	21.64	13.60
19	Manufacture of leather goods	35.70	43.41	22.82	16.37
20	Manufacture of wood and wood products	8.96	14.04	11.41	2.34
21	Paper and paper products	15.32	10.91	8.56	15.87
22	Printing and publishing	2.47	1.44	(0.45)	9.32
23	Coke and refined petroleum products	86.57	21.81	(22.17)	(3.62)
24	Chemicals and chemical products	14.58	29.78	31.23	12.75
25	Rubber and plastic products	13.81	24.11	34.32	19.79
26	Nonmetallic mineral products	4.90	22.96	49.38	11.46
27	Manufacture of basic metals	29.57	44.96	20.45	7.81
28	Fabricated metal products	11.53	24.76	35.16	15.77
29	Machinery and equipment	10.89	36.34	41.76	9.60
30	Office accounting and computing machinery	2.87	96.96	210.07	(3.35)
31	Electrical machinery and apparatus	13.86	34.52	46.63	20.23
32	Radio TV and communication equipment	45.96	81.42	26.71	(3.59)
33	Medical and optical instruments and watches and clocks	20.63	74.38	13.88	13.88
34	Motor vehicles, trailers, semi-trailers	70.64	80.48	19.66	9.40
35	Manufacture of other transport equipment	45.04	75.68	35.09	11.80
36	Manufacture of furniture manufacturing	7.53	28.63	43.98	10.36
37	Recycling	0.00	0.00	0.00	0.00
	Total IT goods (30 + 32)	45.38	83.48	30.25	(3.61)
	Total manufacturing	18.12	29.56	23.84	11.31

Source: Estimates based on data presented in General Statistical Office, *Statistical Yearbook 2001*, Statistical Publishing House, Hanoi.

This is in tune with the experience of Spain in the first years of European Union (EU) membership and Mexico prior to North American Free Trade Agreement (NAFTA) entry. They also show that FDI inflows are expected to double, and the annual GDP growth is likely to be about 0.6 per cent points higher than without BTA.

Trade policy reforms

During the last decade significant changes have been initiated with a view to create more liberal trade regime.[23] These changes related not only to reduction in the taxes and tariffs on imports and exports, but also to the freedom for enterprises to engage in trade. Earlier, only those licensed by the Ministry of Trade were allowed to engage in trade. With the abolition of import–export licensing system in 1998 all the Vietnamese-owned enterprises, irrespective of their ownership structure, were allowed to import and export goods. Through the Decree No. 10/1998/ND-CP dated 23 January 1998, the government allowed the foreign-invested manufacturing enterprises to purchase goods in the local market for direct export or export after processing. The freeing-up of trading rights has prompted rapid growth in the number of enterprises that export and/or import. Nearly 3000 additional private firms sought custom-codes within the year of 1999 after freeing trading rights.[24]

The tariff schedule has been subjected to a series of changes over time in tune with changing policy regime in the country. However, the policy reforms have been severely constrained by the pressures to protect the less efficient state-owned industries which are often overemployed. Nevertheless, Vietnam is committed to reducing or eliminating tariffs and other trade restrictions, since it is a requirement for its membership into AFTA and if it is to realize its hopes for membership in the WTO. The bilateral trade agreement between the US and Vietnam also addresses various market access considerations, including both tariff and non-tariff barriers. Thus the government is sandwiched between the domestic forces for import substitution and external pressures for opening up.

Yet the available data tend to indicate that both nominal and effective rate of protection have fallen (Figure 7.1). But a disaggregate analysis by the Institute of Economics (2002) shows that between 1997 and 2002 the effective rates of protection for export-oriented goods, including agricultural products, have increased even though the nominal tariff for these products decreased. In case of electrical and electronics, the estimated protection coefficients are found to be over 65 per cent. It has also been shown that both domestic and foreign investments in Vietnam are being directed towards sectors with relatively high levels

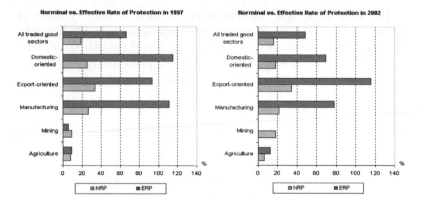

Figure 7.1   Nominal and effective protection rates by sectors in Vietnam: 1997 and 2002
Source: World Bank (2003).

of protection and not towards sectors that are viable with low levels of protection. To elaborate further, about 50 per cent of investment has been in sectors with effective rates of protection over 90 per cent, and a quarter is in sectors with effective rates over 120 per cent and only about 14 per cent of the value of FDI is in sectors with effective rates less than or equal to zero – most of which being oil or gas projects (Center for International Economics 1999).

While tariff and non-tariff barriers set limits to imports in a number of sectors of the economy, the policy reforms initiated in the IT and software sector have had the effect of making IT fairly free from such restrictions. Our estimates using UNCTAD data showed that in 2001, Vietnam imported IT goods worth $636 million accounting for about 4 per cent of the total imports into the country in 2001. The imports originated from about 46 countries, of which 11 countries led by Singapore (35 per cent) and Japan (13 per cent) accounted for about 93 per cent of the total imports. It was also noted that Vietnam has a highly diversified import structure wherein no single product dominates.

IV. Production and use of IT

Present state of IT production

Software

The policy initiatives and the institutional interventions by the government have begun to bear fruits. Estimates from the software industry sources reveal that the value of Vietnam's ICT industry (covering software, hardware, network services and systems integration) has increased from about $337 million in 2000 to US$417 in 2002 and $690 in 2003. Surveys conducted by the International Data group indicate that the IT market in Vietnam has been recording an annual growth rate of 25 per cent and this rate of growth is expected to continue through 2010.[25] Another survey by the PC world has shown that by the end of 2002 there were about 260 software companies, employing about 5000 specialists and the recorded growth rates in sales were of the order of 23 per cent. On an average, these firms employ 20 persons and the turnover per person for the year 2000–2001 was $6400 and $11,000 for software companies and software outsourcing companies respectively. The recorded sales per employee are much lower than those of leading players like India wherein the sales per employee has been of the order US$16,000 in 1999 (Parthasarathi and Joseph 2004). In general the revenue per employee depends to a great extent on the nature of software-related activities undertaken by the firms, which in turn depends *inter alia* on the availability of skilled manpower and the telecommunication infrastructure and its costs. The revenue per employee is likely to be higher if the firms are engaged in software products. However, given the relatively low level of skill, hardly any firms in Vietnam are found engaged in software products. In such a situation, the firms are more likely to be working on relatively low skill activities like coding and maintenance wherein the revenue per employee is likely to be low. Similarly, the weak telecom infrastructure coupled with high cost might induce the firms to resort to onsite development, which also might contribute towards low revenue per employee.

Given the fact that there are a large number of non-resident Vietnamese working in foreign countries and are interested in participating in the IT activities in the country, more concrete steps are to be initiated to utilize this potential for the development of the IT sector in the country. According to one report, there are about 50,000 Vietnamese who play important role in the IT across the world. In the Silicon Valley alone, there are about 10,000 Vietnamese who work in the field of IT (TuyHoa 2001). Towards attracting these people into Vietnam, much could be learned from the Indian and Chinese experience. In the late 1990s Chinese policy makers, academic institutions and technology companies increased their commitment to improving external communications, particularly with the overseas Chinese in Silicon Valley. They sponsored an increasing number of events and programs in the US, while also inviting Overseas Chinese academics and industry representatives to China to attend conferences and other events. In addition, the Ministry of Education established the "Chunhui Program" to finance short-term trips to China by Overseas Chinese to participate in technology-related programs. The result has been that 140,000 students trained in the US returned to China during 1996–2000 and started 3000 firms with a total output of US$1 billion (Saxenian 2002).

Hardware and electronics

It is understood from discussion with the electronics industry association that about 4,00,000 computers have been produced in the year 2002 and the industry has been recording a growth rate of 30–40 per cent in the recent years. About 60 per cent of this has been accounted by the non-branded sector. With a view to gain a better understanding of the growth of IT goods sectors, we have analyzed the data on the production of IT goods sector (TV, radio, communication equipment-ICIC 32, computing machines) and compared its share as well as growth of the industrial sector as a whole and other industries (two-digit level). The results of such an exercise is presented in Table 7.6.

The analysis presented in the table refers to the period 1995–2001 for which published data was available. To begin with, it may be noted that the industry segment office accounting and computing machinery recorded a very high growth rate of over 72 per cent, *albeit* from a very low base, which is in tune with the observations made by the industry associations. When it comes to the industry segment, radio, TV

Table 7.6 Structure and growth of manufacturing output in Vietnam (bill dongs at 1994 prices)

ISIC	Industries	1995		2001		ACGR
		Output	Share	Output	Share	
15	Food products and beverages	27,008.2	32.44	49,388.6	26.83	10.58
16	Manufacture of tobacco products	3,976.7	4.78	6,635.7	3.60	8.91
17	Manufacture of textiles	6,176.2	7.42	11,577.6	6.29	11.04
18	Manufacture of wearing apparel	2,949.8	3.54	6,923.5	3.76	15.28
19	Manufacture of leather goods	3,569.9	4.29	10,074.4	5.47	18.88
20	Manufacture of wood and wood products	3,323.5	3.99	4,045.6	2.20	3.33

21	Paper and paper products	1,946.8	2.34	4,478.8	2.43	14.90
22	Printing and publishing	1,510.4	1.81	2,551.9	1.39	9.13
23	Coke and refined petroleum products	343.2	0.41	305.3	0.17	(1.93)
24	Chemicals and chemical products	5,085.6	6.11	12,713.8	6.91	16.51
25	Rubber and plastic products	2,272	2.73	7,624.1	4.14	22.35
26	Nonmetallic mineral products	9,200	11.05	21,781.3	11.83	15.44
27	Manufacture of basic metals	3,427.9	4.12	6,885.8	3.74	12.33
28	Fabricated metal products	2,331.6	2.80	6,601.2	3.59	18.94
29	Machinery and equipment	1,345.1	1.62	3,263.3	1.77	15.91
30	*Office accounting and computing machinery*	*27.9*	*0.03*	*733.4*	*0.40*	*72.43*
31	Electrical machinery and apparatus	1,087.6	1.31	4,319.6	2.35	25.85
32	*Radio, TV and communication equipment*	*2,064.8*	*2.48*	*4,818.7*	*2.62*	*15.17*
33	Medical and optical instruments and watches and clocks	202.6	0.24	469.5	0.26	15.03
34	Motor vehicles, trailers, semi-trailers	1,459.7	1.75	3,762	2.04	17.10
35	Manufacture of other transport equipment	1,892.7	2.27	8,353.5	4.54	28.08
36	Manufacture of furniture manufacturing	1,969.5	2.37	4,611	2.50	15.23
37	Recycling	88.8	0.11	172.8	0.09	11.65
	Total IT goods (30 + 32)	*2,092.7*	*2.51*	*5,552.1*	*3.02*	*17.65*
	Total manufacturing	83,260.5	100.00	184,092.4	100.00	14.13

Note: Value of output is given in billion Dongs. ACGR – Annual Compound Growth Rate
Source: Estimates based on data presented in General Statistical Office, *Statistical Yearbook 2001*, Statistical Publishing house, Hanoi.

and communication equipment, the recorded growth rate has been only a little over 15 per cent, only marginally higher than the growth rate recorded by the manufacturing sector as a whole (14.13 per cent). Incidentally, there were a number of industries which recorded higher growth rate than this industry. If we look at the IT sector as a whole, its share in the manufacturing output increased only by about 0.5 per cent (from 2.5 per cent in 1995 to 3.0 per cent in 2002) and the recorded growth rate has been only of the order of 17.6 per cent. Thus, it appears that there is the need for greater focus on developing the IT goods sector, which could not only be a source of income and employment in the domestic economy but also a source of foreign exchange.

IT use: Telephones (fixed)

Even with a meager per capita income level, Vietnam passed the critical threshold point of one fixed line per 100 inhabitants (teledensity) as early as in 1994, some two years after Indonesia (ITU 2002d).[26] Since 1995 the telecom lines in the country has grown in leaps and bounds (Table 7.7). The recorded growth rate during 1995–2000 for the country as a whole was as high as over 31 per cent. What is more important is that the growth has been almost uniformly distributed across different regions. From discussion with the senior officials it was discerned that by December 2002 the fixed telephone lines in the country has increased to 3664, 752 recording a teledensity of 6.9 per cent – almost sevenfold increase as compared to 1994.

Experience of other countries in the region tend to suggest that with increase in telephone access there has also been significant increase in the inter-regional variation in teledensity leading to what is called

Table 7.7 Growth of fixed telephone lines in Vietnam

Region	Number of Lines		Annual Compound Growth Rate
	1995	2000	
Red River delta	203,874	778,515	30.7
Northeast	48,385	179,549	30.0
Northwest	7,490	26,322	28.6
North Central coast	43,947	185,107	33.3
South Central coast	58,030	213,108	29.7
Central highlands	31,286	110,649	28.7
Southeast	238,308	996,272	33.1
Mekong River delta	103,035	414,654	32.1
Others	12,112	NA	
Whole country	746,467	2,904,146	31.2

Source: Estimates based on data presented in General Statistical Office, *Statistical Yearbook 2001*, Statistical Publishing house, Hanoi.

the "intra-national telecom divide". Hence it may be instructive to examine how Vietnam fared in this respect under the state monopoly. Table 7.8 presents inter-regional variation in teledensity during 1995–2000. It may be noted that the inter-regional variation across different provinces declined substantially (see the estimated Coefficient of Variation presented in Table 7.8) during the five-year period under consideration wherein the number of telephones in the country recorded more than 30 per cent ACGR. More importantly, the observed decline has been almost across different regions in the country. The only exception has been the North Central region wherein there has been substantial increase in the intra-regional variation. The other two regions wherein there was a marginal increase in intra-regional variation are the Red River Delta and the North West region.

IT use: Mobile telephone

According to ITU (2002d), by December 2000 there were about 7,88,559 mobile subscribers in the country. Within a short span of two years (by December 2002) the number of mobile subscribers has more than doubled to reach a level of 1.9 million and about 2.5 million by April 2003. There are mainly two GSM-based operators in the mobile sector, Vinaphone and Mobifone.[27] While Mobifone with a market share of about 45 per cent caters mainly to the private sector, Vinafone with a market share of 54 per cent is having its customer base mostly with the government.

IT use: Internet and computers

The number of Internet subscribers in the country increased over time from about 1,00,000 in 2000 to 1,96,416 by December 2002. Given the growing number of Internet cafes (about 3500) offering Internet services at an affordable cost of VND6000/hour, the actual user base is probably 2–3 times that number. However, this represents about 0.7 per cent of the country's 80 million inhabitants – far below the regional Internet penetration average of 1.1 per cent. While the number of Internet users has shown remarkable increase, one estimate shows that the actual Internet use by Vietnamese is rather low. It was shown that about 50 per cent of the Internet users in Vietnam are foreigners in Vietnam, 27 per cent are enterprises and only the remainder (23 per cent) is private Vietnamese users.[28] Given the fact that Internet cafes are limited to urban centers like Hanoi, HCMC and Danang, it might be leading to a situation wherein Internet access is confined mostly to the privileged

Table 7.8 Inter-regional variation in growth of telephones and teledensity in Vietnam

Province	1995			2000		
	Telephones	Population	Teledensity	Telephones	Population	Teledensity
Red River Delta	*203,874*	*161,367*	*1.26*	*778,515*	*170,392*	*4.57*
Hanoi	138,051	24,310	5.68	471,846	27,392	17.23
Hai Phong	18,260	16,082	1.14	83,877	16,944	4.95
Vinh Phuc	1,781	10,482	0.17	188,446	11,059	17.04
Ha tay	10,199	22,990	0.44	47,420	24,141	1.96
Bacninh	3,895	9,160	0.43	26,301	9,488	2.77
Hai Duong	9,554	16,091	0.59	34,269	16,631	2.06
Hung Yen	2,474	10,332	0.24	16,266	10,805	1.51
Hanam	1,915	7,637	0.25	12,504	7,955	1.57
Nam Dinh	8,027	18,205	0.44	32,797	19,041	1.72
Thai Binh	7,000	17,523	0.40	21,859	18,038	1.21
Ninh Binh	2,718	8,555	0.32	12,930	8,898	1.45
SD			1.60			6.15
CV			126.66			134.68
North east region	*48,385*	*83,989*	*0.58*	*179,549*	*84,428*	*2.13*
Ha Giang	2,150	5,503	0.39	6,947	6,166	1.13
Cao Bang	1,980	4,892	0.40	7,105	4,965	1.43
Lao Cai	2,850	5,501	0.52	11,110	6,071	1.83
Bac Kan	424	2,542	0.17	3,689	2,801	1.32
Lang Som	4,440	6,792	0.65	19,417	7,123	2.73
Tuyen Quang	2,240	6,388	0.35	8,275	6,840	1.21
Yen Bai	2,988	6,477	0.46	10,212	6,907	1.48
Tahi Nguyen	6,201	1,050	5.91	18,672	10,545	1.77

Phu Tho	5,594	12,117	0.46	21,528	12,746	1.69
Bac Giang	5,543	14,310	0.39	20,712	15,104	1.37
Quang Ninh	13,975	9,417	1.48	51,882	10,160	5.11
SD			1.66			1.15
CV			287.58			53.86
North west	*7,490*	*20,657*	*0.36*	*26,322*	*22,780*	*1.16*
Lai Chau	2,116	5,355	0.40	6,249	6,043	1.03
Son la	2,726	8,117	0.34	9,831	9,059	1.09
Hoa Binh	2,648	7,185	0.37	10,242	7,678	1.33
SD			0.03			0.16
CV			8.19			13.88
North central coast	*43,947*	*95,806*	*0.46*	*185,107*	*101,018*	*1.83*
Thanh Hoa	7,065	33,377	0.21	38,971	34,940	1.12
Nghe An	14,805	27,149	0.55	64,451	28,871	2.23
Hatinh	3,965	12,477	0.32	15,257	1,275	11.97
Quang Binh	3,820	7,461	0.51	15,096	8,016	1.88
quang tri	4,986	5,349	0.93	15,737	5,806	2.71
Thua Thien–Hue	9,306	9,993	0.93	35,595	10,635	3.35
SD			0.30			4.03
CV			65.94			220.17
South central coast	*58,030*	*62,024*	*0.94*	*213,108*	*66,254*	*3.22*
Da Nang	17,000	6,373	2.67	61,429	7,035	8.73

Table 7.8 (Continued)

Province	1995			2000		
	Telephones	Population	Teledensity	Telephones	Population	Teledensity
Quang Nam	4,878	13,220	0.37	19,641	13,894	1.41
Quang Ngai	8,226	11,490	0.72	24,631	12,001	2.05
Binh dinh	10,200	13,944	0.73	36,357	14,810	2.45
Phu Yen	4,663	7,403	0.63	15,618	8,007	1.95
Khanh Hoa	14,194	9,594	1.48	42,058	10,507	4.00
SD			0.85			2.74
CV			91.19			85.19
Central highlands	*31,286*	*33,848*	*0.92*	*110,649*	*42,367*	*2.61*
Kon Tum	2,063	2,795	0.74	7,292	3,248	2.25
Gialai	5,435	8,507	0.64	22,580	10,170	2.22
Dak Lak	9,594	13,983	0.69	38,719	18,609	2.08
Lam dong	14,194	8,563	1.66	42,058	10,340	4.07
SD			0.49			0.95
CV			52.65			36.20
South east	*238,308*	*106,945*	*2.23*	*996,272*	*120,668*	*8.26*
HCMC	175,106	46,404	3.77	699,760	52,261	13.39
Ninh Thuan	3,665	4,665	0.79	15,808	55,148	0.29
Binh Phuoc	2,949	5,332	0.55	13,840	6,846	2.02
Tay Ninh	8,606	9,100	0.95	33,323	9,763	3.41
Binh Duong	10,420	6,390	1.63	42,355	7,377	5.74

Dong Nai	17,471	18,448	0.95	98,207	20,394	4.82
Binh Thuan	8,933	9,517	0.94	33,140	10,659	3.11
Ba Ria Vung Tau	11,158	7,089	1.57	59,839	8,220	7.28
SD			1.03			1.03
CV			46.25			12.48
Mekong River Delta	*103,035*	*155,319*	*0.66*	*414,654*	*163,447*	*2.54*
Long an	9,405	12,508	0.75	31,895	13,303	2.40
Dong Thap	8,252	14,893	0.55	31,297	15,782	1.98
An Giang	13,293	19,701	0.67	55,999	20,770	2.70
Tien Giang	9,074	15,815	0.57	37,748	16,230	2.33
Vinh Long	5,677	9,904	0.57	25,391	10,177	2.49
Ben tre	6,175	12,818	0.48	29,890	13,054	2.29
Kien Giang	12,117	13,920	0.87	43,921	15,240	2.88
Can Tho	14,532	17,397	0.84	62,115	18,362	3.38
Tra Vinh	5,536	9,349	0.59	21,029	9,783	2.15
Soc Trang	6,560	11,501	0.57	24,434	11,910	2.05
Bac lieu	9,414	7,094	1.33	21,468	7,443	2.88
Ca mau	3,000	10,418	0.29	29,467	11,393	2.59
SD			0.26			0.26
CV			38.97			10.19
Total	*746,467*	*719,955*	*1.04*	*2,904,176*	*776,354*	*3.74*
SD: all provinces			18.08			31.92
CV: all provinces (%)			1,743.85			853.24

Source: Estimates based on data presented in General Statistical Office, *Statistical Yearbook 2001*, Statistical Publishing house, Hanoi.

urban population. If we assume that about 70 per cent of the Vietnamese users are in the urban areas, then, one is likely to conclude that much needs to be done to make available the fruits of new technology to the large mass of population living in the rural and peri-urban areas.

While the policy underscores the need for software development for exports, the cost of infrastructure appears to be extremely high as compared to the neighboring countries. Broadband options are currently limited to leased high-speed circuits ranging in speed from 64 kbps to 2.048 Mbps. In 2003, cost of a 64 kbps leased line was around $1442 per month while 2 Mbps line $1734 per month.[29]

It goes without saying that the enviable telecom success that Vietnam achieved is unlikely to be replicated in the case of Internet unless more imaginative steps are undertaken. To begin with, there appears to be the need to de-concentrate the Internet market and attract more investment. Secondly, it is important to explore if the policies have given due attention to use IT for addressing the issues of agricultural sector, especially the commercial agriculture which is currently facing unprecedented crisis. Here the government has been proactive and initiated a few imaginative steps. In its attempt to provide improved access to the provinces and communes, a number of pilot projects to extend Internet to post offices and community centers have already been initiated. Also in its attempt to enhancing the Internet access in rural areas, a plan to widen Internet use at commune level has been underway. It is also planned to set up 400 cultural post points with free Internet access to deliver useful information on agricultural production, cultivation, husbandry and so on to address farmer's information needs. It is scheduled to carry out a preliminary assessment of this pilot activity before taking further steps forward to accomplish this plan and decide to expand it to 5000 points of this type over the country in the forthcoming years (USAID 2001). Yet it may be borne in mind that given the complementarity between Internet access and Internet content on the one hand and the capacity to make effective use of the content on the other, any lopsided approach towards simply providing Internet access may not bring the desired outcome. In this regard, there is the need for bringing together different stakeholders like private sector, NGOs, provincial government and so on.

IT use in government[30]

In 1990, the Ministers Council's Standing Board ratified the project on the application of Informatics and information techniques at the government offices. The project aimed at computerizing the management information system at the government office and at the same time providing partial support in terms of equipment, technology and personal training for ten ministries and ten key provinces. After promulgating the resolution (No. 49/CP) on IT development dated August 1993, the prime minister ratified the five-year plan on the deployment of the National IT program. During 1995–98 period around 50 per cent of the program fund (VND 160 million) was spent on computerizing the state management information system. By 2003 the system of LAN has been set up in 61 provinces and centrally run cities and almost all the ministries and the ministerial level agencies attached to the government. The government's Wide Area Network (CAPNET)[31] was established to link the central computer network of 61 provinces and cities with around 40 key agencies of the government through 2500 terminals and 180 servers nationwide and 50 different applied programs. In addition to the government's WAN, six national databases have been formulated, including the national database on finance, socio-economic statistics, laws, land resources and population. Computer training has been provided along with building up of the computer network.

V. Concluding observations

Ever since the initiation of policies for renovation (Doi Moi) in the mid-1990s, the economy of Vietnam witnessed unprecedented growth and structural change and also managed to traverse almost successfully through the troubled weather created by Asian crisis. No wonder, commentators have rightly described Vietnam as another "tiger" in the making. Along with the shift towards market-oriented policies and greater integration with the rest of the world, the Government of Vietnam also has undertaken measures to build up a vibrant innovation system, which in turn seem to have given rich dividends.

The study observed that with respect the telecommunication (landline) Vietnam has been able to record enviable growth. More importantly, such a high growth has been distributed almost uniformly across the region such that there has been a significant reduction in the intra-national telecom divides. In this respect, performance of Vietnam appeared to be much better than other ASEAN New Comers. But when it came to Internet, the performance appeared to be less remarkable. Not only that the cost of Internet is high, the access is limited mostly to the urban rich and much needs to be done to make available the fruits of this new technology to the large mass of population living in the rural and semi-urban areas. In this regard, there appears to be the need for de-concentrating the industry, which in turn is likely to facilitate more investment in building up IT infrastructure.

Vietnam today has a small, yet thriving, IT sector including both hardware and software. The observed rate of growth in both hardware and software production is found almost on par with the target set by the policy and also much better compared to other ASEAN New Comers. The various ingenious initiatives made by the government, including the setting up of different STPs are likely to give rich dividends in the near future. Yet the present level of revenue per employee earned by the software firms is at a very low level reflecting the nature of software activities undertaken by the firms. To the extent that the firms are currently operating at the lower end of the value chain, concerted efforts towards further strengthening the innovation system coupled with cost-effective and quality IT infrastructure and quality manpower are called for such that the firms move up the value chain. The study also underscores the need for initiating measures to promote investment from the Overseas Vietnamese. In this respect much could be learned from India and China.

While ambitious targets coupled with concerted actions have been made towards developing IT manpower, the focus so far appears to have been on quantity. The outcome appears to have been the creation of a large pool of "unemployable" human power, which is likely to be a liability rather than an asset unless the production structure of IT is planned accordingly. Hence there appears to be the need for deliberate policies to improve the quality of IT training which would involve accreditation for training institutions and the courses, developing strong linkages between the academia and the industry, developing partnership with countries like India and other appropriate measures.

Given the fact that affordability is a major issue in promoting the use of IT in developing countries like Vietnam, which arises on account of the high price of hardware and software in relation to the income levels, there appears to be the need for greater focus on promoting the use of open source software. However, open source software is yet to receive the attention of policy makers that it deserves in a country like Vietnam. It also appears that the issue of developing local content for addressing issues specific to Vietnam, especially the rural/agricultural sector, needs further attention. This is not to undermine the various efforts made towards providing Internet access in the rural areas. Yet the study underscores the complementarities between Internet access and Internet content on the one hand and ability to make effective use of the content on the other. In such a context, any lopsided approach towards simply providing costly IT infrastructure has the potential danger of uneconomic use of scarce resources and not obtaining the desired outcome. In this regard, there is the need for bringing together different stakeholders like private sector, NGOs, provincial government and so on. The various policy initiatives and institutional interventions seem to have overlooked the role of some of these actors.

Various policy reforms to liberalize the trade regime notwithstanding, the study notes high levels of effective rate of protection across different industries. This in turn has had its adverse impact on the cost and competitiveness of sectors like agriculture. Moreover, it was also noted that both foreign and local investments have increasingly been concentrated in sectors with higher levels of protection. In this respect, IT was no exception. To the extent that FDI has been influenced by higher protection, further trade liberalization might be instrumental in creating a more competitive environment. At the same time, if such trade liberalization is likely to have adverse impact on the state-owned enterprises, the process of trade liberalization might be undertaken in phased manner along with concerted efforts towards enhancing their competitiveness through appropriate policy initiatives.

8
ICT and the Developing Countries: Towards a Way Forward

I. Introduction

Today there is a widely held consensus on the potential of ICT to contribute towards socio-economic transformation of the developing countries. This has induced unparalleled initiatives undertaken at the instance of not only governments – both national and sub-national – but also the NGOs and Multilateral Organizations towards harnessing this technology for development. The initiatives, in general, lay emphasis on trade and investment liberalization and place the developing countries in a situation of perpetual *attente* – waiting for the transfers of technology from the North and focusing their attention on the need to attract transnational corporations to their shore (Mytelka and Ohiorhenuan 2000). Such an approach ignores the technological capabilities built up in the South over the past decades and has the danger of perpetuating technological dependence of the South (Joseph 2005b). No wonder scholars (e.g. Parayil 2005) have been concerned with the current digital world order and the discussions throughout the World Summit on Information Society (WSIS) process made it clear that a majority of developing countries were unsatisfied with the status quo in this matter and called for a change (UNCTAD 2004b).

In this context, this book sought answers to certain issues with a view to contribute towards informed policy making in developing countries aspiring to catch up with the ongoing ICT revolution. The issues centered on the relative role of trade regime and innovation system in enabling the development of IT sector in developing countries. Seeking answers to the issues involved an exploration of the factors that configured the ICT success of a developing country like India and the present state of laggardness of many developing countries that since long followed a liberal trade and investment policy regime. At the operational level, the study involved an analysis of the present state of innovation system, trade and investment regime and achievements with respect to ICT production and use in India and five ASEAN countries. Among these countries, Thailand represented countries with a longer history of liberal trade and investment policies (old ASEAN) and the other four (Cambodia, Laos, Myanmar and Vietnam) are the ASEAN New Comers that recently resorted to a liberalized policy regime. Given the fact that the countries studied represent the broad cross-section of developing countries in terms of development strategies adopted and levels of development achieved, the conclusions and inferences drawn may be of relevance not only to India and ASEAN, but also to the developing countries in general.

Analytically, the present study centered on the following premises: First, the developing countries could gain from both the production and the use of ICT as much as, if not more than, their counterparts in the developed world. Secondly, there is a link between trade and investment in general and this relationship is much stronger in case of IT. Hence, if the production, and therefore investment, in ICT is to take place in any economy, the trade policy regime needs to be the one wherein the free flow of inputs into and outputs out of the economy is ensured. Thirdly, trade policies, given their bearing on the access to and price of ICT goods and services, could be instrumental in influencing the use of ICT, especially in countries where affordability is a central issue on account of the high price of ICT goods and services. Fourthly, domestic production capability of ICT goods and services could act as a catalyst in ICT diffusion through easy access on the one hand and the generation of income-earning opportunities leading to higher demand on the other. Trade and investment liberalization, however, is only a necessary condition and therefore a country's success in promoting production and use of ICT depends on its success in building up an innovation system. This *inter alia* calls for institutional arrangements at the instance of different stakeholders and the interactions among them to facilitating the production and use of ICT. Finally, given the capabilities that exist in select developing countries, there appears to be an unexplored avenue to harness the SIS to hasten the catching-up process by building new bridges in South–South cooperation.

This concluding chapter highlights the main findings of the country case studies to draw the broad contours of a plausible way forward that may be of relevance to developing countries at large. At the same time, given the bearing of ongoing multilateral and regional initiatives in ICT in developing countries, we shall also indicate the plausible future directions of change in such initiatives to serve better the interests of developing countries. Finally, a case has been made for an e-South Framework Agreement encompassing the complementarity between trade regime and innovation system and harnessing the SIS to hasten the process of catching up by developing countries.

II. India and Thailand

It was shown that among developing countries India stands head and shoulders above most other developing countries in terms of her achievements in developing an internationally competitive IT software and service industry. Of late, India has also emerged as the most attractive location among the developing countries for the outsourcing of IT-enabled services. Moreover, India can boast of a number of e-governance projects and various ICT programmes to address development issues in different sectors undertaken at the instance of different stakeholders though her success in promoting ICT use in general has been less remarkable as compared to the performance in export of ICT software and services. Nonetheless, India's overall performance in terms of her rank in NRI has been found comparable to those countries having per capita income three to four times higher than India. The remarkable overall performance notwithstanding, the study underscores the need for greater focus on the domestic market to further accelerate the diffusion of ICT into different sectors of the economy.

Our analysis of the contributory factors towards the emergence of India as an ICT powerhouse from the south has shown that the NSI, which evolved over time at the instance of state policies and strategies, has been instrumental in facilitating it. The Indian Government recognized the potential of the country in computer software way back in the early 1970s and undertook various initiatives. These initiatives included the development of a system of higher education in engineering and technical disciplines, creation of an institutional infrastructure for S&T policy making and its implementation, building centers of excellence and numerous other institutions for technology development. Along with building the skilled manpower base, government also addressed some of the infrastructural constraints by setting up a number of STPs in different parts of the country. The government also facilitated technological-capability building with investments in public-funded R&D institutions and supporting their projects, by creating computing facilities, and developing an information infrastructure for data transfer and networking.

In a sense the returns to the innovation system built up over the years remained at best modest until the trade and investment regime was liberalized, indeed at her own terms and at her own pace, and assigned greater role for the private sector, industry associations and other stakeholders like the NGOs and different state governments. Thus India's IT success is a typical case of proactive state intervention wherein the government laid the foundation and created the facilitating environment and the industry took off under the liberal trade regime and increased world demand in the 1990s *inter alia* on account of the Y2K problem. Thus viewed, Indian experience highlights the complimentary role of trade and investment policies and innovation system in facilitating the production and use of ICT in developing countries.

The case study of Thailand further reinforced the inference drawn from the Indian experience. It is found that liberal trade and investment policies followed by Thailand for a longer period has been instrumental in achieving higher output growth along with structural transformation wherein the industrial sector, and more specifically the manufacturing sector, emerged as the most vibrant sector of the economy. The liberal policy regime coupled with other facilitating environments like good infrastructure and the abundant supply of cheap labor led to substantial investment in electronics making it a major source of employment and export earning. However, the study finds that electronics industry in Thailand has been characterized by sticky specialization in a few low technology products leading to low value addition, poor forward–backward linkages and getting locked up in the low end of the electronics value chain. More importantly, despite the liberal trade and investment policy regime, Thailand has not been able to make its presence felt in the highly competitive and skill-intensive areas like IT software and service sector.

The differential performance of Thailand with respect to IT as compared to India, despite its liberal trade regime, has been attributed to its neglect in fostering a vibrant NSI during the early years. This has led to limited skill and knowledge base, which in turn acted as a stumbling block for the establishment of high value-adding skill-intensive activities like IT software and services. What is more, while encouraging investment, the incentive structure was not tuned to induce the companies, both foreign and local, to invest in innovation, skill upgradation and knowledge generation. Thus the single most important hurdle that Thailand faces in making headway in the sphere of IT is the scarcity of human capital, in terms of both quantity and quality. Despite various initiatives, especially during the post-crisis period, the IT sector of Thailand is faced with an excess demand situation for IT manpower. Unlike India, which moved towards a liberal trade and investment regime after building up an innovation system, Thailand, of late, has been making earnest attempts towards building up an innovation system, especially through its IT Master Plan (2002–2006) with highly ambitious targets in terms of ICT production and use. The IT Master Plan involves various institutional interventions and policy measures for strengthening skill base, promoting R&D investment and fostering an interface between academia and industry. Also, there is evidence to suggest a greater orientation in the private and public sectors towards innovative activities. Such a shift from investment-led growth to innovation-driven growth strategy adopted in the recent years is expected to give rich dividends in the near future.

Now the key issue is the extent to which the ASEAN New Comers have imbibed lessons from the experience of other countries like India and Thailand. An answer to this issue may be obtained from our analysis of the present state of innovation system, trade and investment policies and performance and finally their success in promoting the production and use of ICT.

III. The ASEAN New Comers

While the ASEAN New Comers in general were faced with making the difficult choice of "investing in Pentium or in Penicillin", they have undertaken a series of bold initiatives towards developing an ICT base and using new technology for addressing their development needs. Nonetheless, given the gigantic task at hand and the rocky road through which they have to traverse, the destination still remains far away.

Innovation system: A point of neglect?

In case of Cambodia, the present policy towards IT lays emphasis on promoting IT production and use in different sectors of the economy and also for promoting e-governance. To achieve this the policy calls for, among others, the development of IT infrastructure, promoting computer literacy, standardization of Khmer language in computers and greater role for the private sector. In Lao PDR the STEA proposed a four-year plan (1996–2000) which dealt with, among other things, developing an IT infrastructure (including human capital, IT industry base, communication network), promoting IT application (in government, business and industries as well as economy at large) and devising policies for promoting ICT development. Achievements by the year 2000 have fallen short of the targets on account of various reasons which *inter alia* included lack of resources and weak interaction among different actors involved. Thereafter, the country has been hastening slowly and has set up five working groups towards formulating an integrated IT policy.

In case of Myanmar, though the computer science law, which governs the use of computers and Internet in the country, and the recently adopted IT Master Plan have certain laudable objectives, the series of restrictions have undermined the positive effect of various initiatives by the government to promote the use of IT for development. Hence the study highlighted the need for phasing out various restrictions on the use of IT in general and Internet in particular. In case of Vietnam, IT production and use is governed mainly by an eminent Vision Document concerning ICT prepared by the Party and subsequently developed Action Plan by the government. Both these documents are unambiguous in their approach, exhaustive in their coverage and ambitious but realistic in their targets. More importantly, unlike other countries, the IT policy in Vietnam upholds the need to develop an innovation system in tune with the recommendations made by a team of experts based on the detailed analysis of S&T system in Vietnam.

General awareness among the policy makers of the need for developing an IT base notwithstanding, the institutional arrangements also were found varying from one country to another. As in India, seven STPs have been set up in different parts of Vietnam, and a beginning has been made in Myanmar with the setting up of an STP in Rangoon and others are yet to follow the suit. While there is a ministry exclusively for ICT in Vietnam, in other countries ICT issues are handled either by the Ministry of Science, Technology and Environment or an Independent Agency like NIDA in Cambodia. There are also instances of more than one agency dealing ICT-related issues leading to coordination failures. Given the pre-eminent role that the new technology plays today, which cuts across different ministries and administrative departments and the imperative of evolving an innovation system, it may be advisable to have an exclusive Ministry for Information Technology in those countries, which are yet to set up a separate ministry.

The policies in almost all the countries seem to be not assigning an appropriate role for the provincial governments in developing and harnessing ICT for development. We have seen that as of now most of the regional governments in India have their own IT policy to promote the production and use of IT in the respective states. While the role of private and public sectors and the coordinated effort has been under-lined in the policies of all the countries, there are other stakeholders like Civil Society Organizations that could play a very constructive role, especially in addressing the issue of "intra-national digital divide" and harnessing ICT for the rural sector in general and agriculture – the mainstay of all the new ASEAN economies – in particular. Also, Indian experience highlights the important role of industry associations. Hence the study underscores the need for making ICT industry associations in ASEAN New Comers more vibrant and fostering links with their counter-parts in countries like India. Given the fact that affordability is a major issue in promoting the use of IT in these countries, which in turn arise on account of the high price of hardware and software in relation to their income levels, there appears to be the need for greater focus on promoting the use of open source software.

While upholding a strategy of export orientation cum import substi-tution, the so-called "strategy of walking on two legs" for the software development, it is important to highlight lessons from India. Such a strategy has a potential threat that the private sector, which dominates software production, might prefer the export market to the relatively less profitable domestic market. This is particularly because of the fact that software-exporting firms are provided with more incentives as compared to their domestic market-oriented counterparts. Hence, it is the respons-ibility of the state to modulate the behavior of private sector in such a way that the domestic use of software and IT are not sacrificed for foreign exchange. This does not mean that export should be discour-aged. The policy may be one of growth-led export and not export-led growth.

A major issue being confronted by these countries relates to the human capital constraint, perhaps an outcome as well as the cause of the weak innovation system. The study argues that present approach of "training the trainers" adopted by most of the countries has its obvious limits and underscores the need for targeted measures to attract more investment into the field of IT education and training. While ambitious targets coupled with concerted actions have been undertaken towards developing IT manpower by most of the countries, the focus so far appears to have been on "quantity", leading to mushrooming of private training institutions. This has the potential danger of creating a pool of "unemployable" human power. Hence there appears to be the need for an accreditation system as practiced in India. Also there is the need to evolve policies to nurture strong linkages between the academia and the industry. This is likely to expose the teachers to the real-world environment through consultancy and other means, and students get opportunities to take up internship with the private sector and the private sector participates in the teaching and development of the curriculum in the academic institutions ultimately resulting in an overall improvement in the quality of manpower. Here much could be gained by joining hands with countries like India known for its institutional arrangements for bringing out high quality IT manpower.

Trade and investment: Policies and performance

As a result of the series of reforms in policies that govern trade and investment undertaken in the recent past by the ASEAN New Comers, the present policy environment in general is conducive to attract investment. Infact, in tune with the trend in the developing countries, these countries are also engaged in incentive competition among each other to attract more investment. Yet, there has been significant inter-country variation in both the amount of investment approved and the fructification (ratio of actual investment to approved investment) rate. Among these countries, Vietnam is found having a much better record as compared to others not only in terms of investment approvals, but also in terms of investment fructification, leading to one of the highest per capita investment in the world. More importantly the inflow of FDI after reaching its peak in 1996 has shown a declining trend in all the countries except for Vietnam, and a trend reversal is yet to take place. This tends to suggest that, despite liberal policy regime, the weak innovation system in Laos, Cambodia and Myanmar appears to act as a debilitating factor in attracting investment because the location decisions of MNCs today are governed by the presence of complementary capabilities.

Improvement in the investment climate in general notwithstanding, there are a number of issues that need to be addressed for promoting investment in the economy at large and ICT sector in particular. For example, it is found that most of the countries follow the policy of stipulating minimum capital requirement for the fully owned foreign enterprises and/or minimum foreign equity share in case of joint ventures. These policies are likely to have the effect of erecting entry barriers to certain foreign enterprises, especially the small- and medium-sized ones, which could also be major source of technology, market access and other intangible and tangible assets needed by the these economies. Also it has been observed that some of the countries follow the practice of case by case exemptions which may be justified for attracting investment in preferred sectors, but it also has a potential danger of decisions turning

212 *Information Technology, Innovation System and Trade Regime*

out to be whimsical and a lack of transparency in the decision making. Therefore, it is advisable to make such concessions more transparent. Also it may be worth considering the option of "automatic approval" for investments in certain areas subject to certain investment limits. To the extent that the business feels the decision-making process in some of the countries is still "lengthy, opaque, and inconsistent", policy makers might consider an open dialogue with the business sector. Another practice which might deter investment related to the need for depositing a performance guarantee and the refund of the same once 30 per cent of the project is completed. These practices might lead to a greater human interface and thus leading to greater scope for corrupt practices. It is also surprising to note that while these countries are plagued by severe regional imbalance in development, there is hardly any provision in the investment law of some of the countries to attract investment into the less privileged regions.

The study tends to suggest that the domestic investment tends to crowd in foreign investment. Since the domestic private sector, for historical reasons, is in its infancy the study calls for further targeted efforts to develop a domestic private sector. This process could start by reviewing and abolishing policies, if any, that are present today which discriminate the local capital *vis a vis* their foreign counterparts. Given the fact the entrepreneurs are "not only born but also developed", special emphasis may be given to create and nurture a domestic entrepreneurial class through EDPs.

Since the present IT production base is limited, the study underscores the need for appropriate policies that in turn will facilitate the establishment of IT production base in the near future. High entry barriers notwithstanding, there might be real opportunities for these countries to enter into those areas of ICT goods characterized by relatively stable technology and low skill requirements like passive components and the assembly of equipment like radio and TV, computers and so on. In the sphere of ICT services, again there appears to be real opportunities to enter into some of the relatively less skill-intensive ICT services like data entry and IT-enabled services like medical transcription and call centers wherein the required skills could be developed in the short run and they are ideal for generating large-scale employment. But such IT-enabled services also calls for better communication infrastructure at affordable prices. In general, the present study underscores the need for initiating steps such that these countries find a place in the international production networks of IT in the near future.

In making efforts towards developing an IT production base, it is important to keep in mind the bearing of innovation system and the lessons offered by the experience of India and Thailand. To begin with, the strategy might be to make available a large pool of IT manpower at different levels such that the primary condition for the establishment of IT goods/service production base is satisfied. Here the strategy needs to be one of pooling together the resources of different actors like the government, private sector and the Civil Society organizations. Also, the strategy should not be one of spreading thinly the resources across the space, instead the investment needs to be undertaken in such a way as to take advantage of the agglomeration economies. This might be possible through the setting up of technology parks wherein built-up space, communication infrastructure and others which are beyond the reach of an individual entrepreneur are provided along with a "single window clearance" system so that the prospective investors need to have only limited interaction with the bureaucracy. Secondly, such technology parks needs to be close to and have constant interaction with the centers of learning such that mutual learning and domestic technological capability is built up in the long run. Thirdly, there is also the need for conscious efforts towards skill empowerment such that the economy does not get locked up in low technology activity and an upward movement along the skill spectrum is ensured.

IT use: Achievements and limits

The study finds that in Cambodia, Lao PDR and Myanmar IT use viewed in terms of telecommunication network (fixed or mobile) and Internet use are confined to the urban areas leading to what is called the "intra-national digital divide". Notwithstanding the competitive environment that the policy upholds to bring about in all the countries, in reality the telecom market is far from competitive *inter alia* because of the dual role that government plays as a provider and regulator of telecom service. But in case of Vietnam, while the government has the monopoly over telephones, the study observed that in telecommunication (landline) the country has been able to record enviable growth since mid-1990s leading to nearly sevenfold increase in teledensity. More importantly, such a high growth has been distributed almost uniformly across different regions such that there has been significant reduction in the intra-national telecom divide. Thus government control *per se* need not necessarily hinder the development of telecom sector. The central issue appears to be the creation of a competitive environment and not necessarily privatization.

The study noted some of the attempts by the Government of Myanmar towards increasing the access to old technologies like TV by increasing transmission coverage and new technology like telephones (both fixed and mobile) and Internet by spending billions of Kyats. But the returns to these investments are yet to be fully reaped. This is evident not only from the low diffusion of old technology like TV and radio, but also from the high intra-national telecom divide and low levels of use of telephones (both fixed and mobile) and Internet. While huge investments were made towards increasing the transmission coverage, low affordability on account of low-income levels coupled with restrictions in the use of TV and radio in the form of licensing seems to have an adverse effect. Similarly, while efforts were made towards modernizing the telecom infrastructure, lack of competition and resultant exorbitantly high prices/costs seem to have had led to the reduced access. Hence the study point towards the need for creating a more competitive environment in the telecom sector in all the countries. But the ways and means of bringing about competition has to be decided by the government and there appears to be no single formula that applies to all.

On the whole, it appears that the overall approach of the ASEAN New Comers so far has been one of making concerted effort at promoting ICT use but the issue of developing a production base appears to have not received the attention that it deserves. This, however, appears to be in tune with the often suggested strategies for the developing countries. Given the fact that such a strategy has the potential danger of perpetuating technological dependence on the one hand and forgoing income-earning and employment opportunities on the other, the present study underscored the need for promoting both production and use. If the experience of India and Thailand is any indication, this calls for simultaneous building up of an innovation system and promoting trade and investment. Given the capabilities that exist in select developing countries, the national policies and strategies need to be tuned to take advantage of the SIS.

Addressing southern problems: Harnessing SIS

Today intra-national digital divide is as acute as international digital divide. In Lao PDR, Cambodia and Myanmar, like in most other developing countries, the capital city accounts for nearly 70 per cent of the total number of telephone lines in the country. Hence, the access to telecommunication and Internet is confined mainly to the urban centers, and the rural areas that accommodate bulk of the population remain beyond the ambit of the new technology. Connecting rural areas is a major challenge because subscribers are geographically dispersed, sparsely populated and economically weak. This makes it unprofitable for the telecom companies to venture into remote villages. Therefore, affordability, ease of deployment and appropriate organizational innovations are critical to sustainable deployment of telecom systems in the countries of the South. Interestingly, there are a number of innovations evolved under the SIS to address these South-specific problems. Such innovations include the CorDECT Wirless in Local Loop Technology Developed in India by the IIT, Chennai.[1] Another technology project that has gone on to pilot stage is DakNet, in the Indian state of Karnataka, where it is attempting to extend the Bhoomi Land Records project to kiosks that do not have connectivity. DakNet offers Wi-Fi-based asynchronous broadband linkage where wired communication is not available (Carlos A. *et al.* 2003).

The issue of affordability arises mainly on account of the high price of hardware and software in relation to the average income level of people in the developing world. Since the pricing strategies of ICT firms are designed mainly by taking into account the developed-world market conditions, the existing business models may not facilitate widespread use of ICT in less developed countries where annual per capita income is only a fraction of their developed-country counterparts. The developments in the FOSS are likely to provide an alternative for the developing world. Several Indian groups are actively at work localizing FOSS to Indian languages, and these include groups like Malayalam Linux, Tamil Linux and others. Even in least developed countries like Lao PDR, attempts are being made to localize the Linux-based graphical desktop and office tools to the Lao language at the instance of Jhai Foundation.[2] In the sphere of hardware, innovations like simputer[3] developed by the Indian Institute of Science, Bangalore, is considered highly relevant in addressing the issues of affordability and illiteracy being faced by developing countries. In this context, the role of South–South cooperation to strengthening SIS by pooling resources to reap economies of scale and scope on the one hand and achieve risk reduction on the other cannot be overemphasized.

From e-ASEAN to e-ASIA

Perhaps, the most notable attempt at harnessing SIS for promoting ICT use and production in the developing world is the e-ASEAN Framework Agreement. The e-ASEAN initiative has to be seen against the background of economic and digital divide between the new ASEAN (Cambodia, Lao PDR, Myanmar and Vietnam) and old ASEAN member countries (Brunei Darussalam, Indonesia, Malaysia, Philippines, Singapore and Thailand). The e-ASEAN initiative, among others, is an integral part of the initiative for ASEAN integration. Hence, unlike the ITA of WTO, which is essentially a tariff-cutting mechanism agreed upon mostly by the developed countries, the e-ASEAN Framework Agreement aims at tariff cutting along with facilitating capacity building (Box 8.1). The underlying strategy is one of "ASEAN help ASEAN".

Here it is pertinent to explore the limits of such a regional arrangement as compared to the benefits of broad-based initiative among countries in the South or at least in Asia. To be more specific, we may examine to what extent the old ASEAN could help in capacity building, both physical and human, in the new ASEAN countries?

Box 8.1   e-ASEAN Framework Agreement

With a view to promote the development of Information communication technology and harnessing it for bridging the development divide between ASEAN member countries, the Heads of ASEAN countries have signed the e-ASEAN Framework Agreement on November 14, 2000 in Singapore.
The specific objectives of this Agreement are to:

(1) promote cooperation to develop, strengthen and enhance the competitiveness of the ICT sector in ASEAN;
(2) promote cooperation to reduce the digital divide within individual ASEAN Member States and amongst ASEAN Member States;
(3) promote cooperation between the public and private sectors in realizing e-ASEAN; and
(4) promote the liberalization of trade in ICT products, ICT services and investments to support the e-ASEAN initiative.

Different measures envisaged in the Agreement are the following.

(a) Facilitating the establishment of the ASEAN Information Infrastructure *inter alia* by enhancing the design and standards of their national information infrastructure with a view to facilitating interconnectivity and ensuring technical interoperability between each other's information infrastructure.

(b) Facilitating the growth of electronic commerce in ASEAN by adopting electronic commerce regulatory and legislative frameworks and other measures that create trust and confidence for consumers and facilitate the transformation of businesses towards the development of e-ASEAN.

(c) Promoting and facilitating the liberalization of trade in ICT products, ICT services and of investments in support of the e-ASEAN initiative by eliminating duties and non-tariff barriers on intra-ASEAN trade in ASEAN ICT products in three tranches. The first tranche shall take effect on 1 January 2003. The second tranche shall take effect on 1 January 2004. The third tranche shall take effect on 1 January 2005. For Cambodia, Lao PDR, Myanmar and Viet Nam, the three tranches are to take effect on 1 January 2008, 2009 and 2010. Also facilitate trade in ICT products by Mutual Recognition Arrangements (MRA) covering ICT products, where applicable, and aligning the national standards to relevant international standards. The agreement also calls for promoting and facilitating investments in the production of ICT products and the provision of ICT services.

(e) Developing an e-Society in ASEAN and capacity building to reduce the digital divide within individual ASEAN Member States and amongst ASEAN Member States by building an e-ASEAN community by promoting awareness, general knowledge and appreciation of ICT, particularly the Internet. In this regard the policy also calls for increasing ICT literacy and expanding the base of ICT workers in the region by regional human resource development programmes covering schools, the community and the work place.

(f) Finally the Agreement calls for promoting the use of ICT applications in the delivery of government services (e-government).

Source: http://www.aseansec.org/5308.htm

If the available evidence is any indication, IT production in the old ASEAN countries to a great extent has been locked up in low value-adding assembly of electronic commodities, and the transition to ODM and OBM status has been rather limited, which in turn point towards limited industrial upgrading. The problem is further compounded by the fact that bulk of the investment in the ICT sector of old ASEAN has been made by the TNCs and therefore these countries have only limited say in the decision to relocate production facilities to new ASEAN member countries. In fact, the old ASEAN countries are well aware of this issue and hence there has been conscious effort during the recent years to move away from the earlier strategy of investment-led growth to one wherein growth is induced by innovation.

Thus there are some limits to ASEAN help ASEAN strategy with respect to IT hardware. This issue is more acute with respect to skill-intensive areas like software. The old ASEAN countries currently faced with an excess demand for IT manpower, in terms of both quality and quantity, need to address this issue for developing a software and service base. Even in case of Singapore, which is more developed among the old ASEAN countries, there is acute shortage of IT manpower. Thus in achieving the declared objective of bridging the development divide between the old and the new ASEAN members by harnessing ICT, cooperation among the ASEAN countries may be complemented with more broad-based cooperation involving other countries in Asia.

At the same time, in the current context wherein the development destiny of developing countries is shaped by their involvement in ITA, the only multilateral initiative, it is important to reflect on its present relevance and the plausible future directions so as to make it more appealing to the developing countries.

Information Technology Agreement of WTO

At the multilateral level, the ITA of WTO, essentially a tariff-cutting mechanism, aims at addressing mainly the issues related to promoting the use and also, to some extent, the production of ICT through the link between trade and investment. However, a preliminary analysis (see Appendix) indicates that even after eight years of its implementation ITA had only 63 signatories, accounting for only less than 50 per cent of 139 WTO member countries. Thus while there is a great rush among developing countries to join WTO and also to harness the new technology for development, they are not generally inclined to sign ITA. While ITA is open to non-members of WTO, all the signatories so far are WTO members. It is observed that during the pre-ITA period (1989–97), the world exports of ITA goods recorded nearly sixfold increase (from $120 billion to $701 billion) with an ACGR of 24.5 per cent. However, the overall performance has been less impressive during the post-ITA period (1997–2003). While the positive growth trends continued since ITA till 2000, though with a lower growth rate (11.5 per cent), the period thereafter

(2000–2003) recorded a negative growth rate (−9.9 per cent). As a result, the level of world exports of ITA goods in 2003 ($710 billion) turned out to be almost at the level that prevailed in 1997 ($701 billion). Also, the developing ITA members as a group performed better as compared to the developed ITA members in terms of exports and imports. But the growth rates in export and import by developing non-ITA countries are found to be much higher than their counterparts who joined ITA.

The empirical evidence, though preliminary, tends to reinforce the inference that emerged from six country case studies presented in this book; trade liberalization *per se* may not be adequate to enabling the developing countries to enjoy the fruits of ICT revolution. Hence ITA would become more attractive to developing countries if the trade liberalization were complimented with targeted efforts at capacity building that is currently missing in ITA. As argued in Chapter 1, the SIS evolved over the years remains an untapped potential in facilitating capacity building as these capabilities are better tuned to the needs of other developing countries.

An e-South Framework Agreement

While a cross-section of countries in the South have built up substantial capabilities in ICT under the SIS, there is a need to further deepen and broaden these capabilities. Also, since IT-enabled services and BPO offer enormous opportunities for generating new income-earning and employment opportunities in the South there is much scope for mutually beneficial cooperation among the Southern countries to learn from each other and to avoid wasteful competition.

Thus the scope for South–South cooperation is too obvious because of IT capabilities in some of the Southern countries and marked divergence in the IT interests of developing and developed countries. But, what is at present missing is an institutional arrangement for promoting the same and research backed by theory and empirics to sustain it. Here lies the need for an e-South Framework Agreement aiming at bridging the digital divide through an integrated development of ICT sector through South–South cooperation. Towards achieving this objective, the Agreement, in tune with the ITA of WTO, should focus on liberalizing the trade in ICT goods and services. At the same time, drawing from the e-ASEAN Framework Agreement the e-South Framework Agreement should be instrumental in capacity building *inter alia* by harnessing the SIS. Given the paramount importance of human capital in developing ICT production and promoting ICT use, special focus may be given on developing IT manpower base wherein there is the need for relaxing the restrictions on the mobility of skilled manpower across the developing world. In general, the Agreement should facilitate an integrated development of the ICT sector by promoting both production and use instead of the ongoing lopsided approach wherein developing countries are often considered as passive adopters of technology. We may conclude by reiterating that South–South cooperation should not be construed as a substitute for the ongoing initiatives at promoting North–South cooperation.

Appendix
ITA: Addressing Digital Divide Through Trade Liberalization

There are many constraints faced by the developing countries in providing ICT access to their population for taking advantage of the benefits of the new technology. The most important among them relates to limited affordability arising from higher prices of ICT goods and services as compared to the income levels in the developing countries.[1] To the extent that a non-competitive environment and the limited ICT infrastructure adds to the problem, there is the need for creating a more competitive environment and attracting substantial investments. But in most developing countries the government budgets are limited and private investment is often deterred by outdated legislation and policies that block investment in new converging technologies. The ITA aims at addressing some of these issues through liberalizing the trade in ICT goods and services.

The ITA, essentially a tariff-cutting mechanism, came into force in 1997 and required the elimination of tariffs and other duties and charges on the goods covered by the ITA in maximal four stages until 2000. However, developing countries could opt for extending their staging until 2005.[2] The participants are required to abide the MFN principle. Hence the benefits of zero tariffs are extended to those WTO members who did not sign the ITA. While ITA is open to non-WTO members, it is not mandatory on the part of WTO members to sign it. Focus of ITA till today is on removing tariff barriers, as the review of non-tariff barriers has not yet come to any conclusion. Also, while the means of mass communication equipment like radio and television could play important role in addressing the information needs of the poor both in developed and in developing countries, such products are not covered by ITA and the negotiations on expanding the product coverage of the ITA have not been concluded. Even today, only 63 countries are signatories of ITA accounting for only less than 50 per cent of 139 WTO member countries. Thus while there is a great rush among developing countries to join WTO and also to harness ICT for development, they are not generally inclined to sign ITA.

Effectiveness of ITA: Preliminary empirical evidence

Different indicators could be employed to assess its effectiveness, such as the number of countries that signed ITA and the extent to which ITA has promoted the demand for IT goods at the global level and across different countries. Given the fact that demand could be met either through trade or through local production, the issue at hand could be analyzed by examining the growth in IT use especially by the developing countries. As the ITA has been fully implemented in a phased manner and full tariff reduction by developing countries have been effective only by 2005 the empirical evidence is only indicative. As we have already seen, with respect to the number of countries joining ITA the picture has not been very encouraging. Now we shall explore to what extent the trade liberalization, as implemented under ITA, has resulted in an increased demand for the IT goods by analyzing the growth in the world trade in ITA goods. Further we shall also explore the relative vibrancy of the IT sector of the developing countries that joined ITA[3] *vis a vis* those did not sign ITA. Since the latter group of countries is highly heterogeneous with wide variation in their levels of development it may be instructive to compare the performance of developing ITA members with non-ITA developing countries with comparable levels of development.[4] Due to the difficulty in obtaining comparable production data across different countries we shall approach this issue by analyzing the export and import of select ICT goods by these countries. Finally we shall reflect on the available evidence of ICT use in developing countries.

Global trade in IT goods

The trend in the world exports of ITA goods during 1989–03 (Figure A.1) shows that during the pre-ITA period (1989–97), the world exports of ITA goods recorded nearly sixfold increase (from $120 to $701 billion) recording an ACGR of 24.5 per cent. However, the overall performance has been less impressive during the post-ITA period (1997–2003). While the positive growth

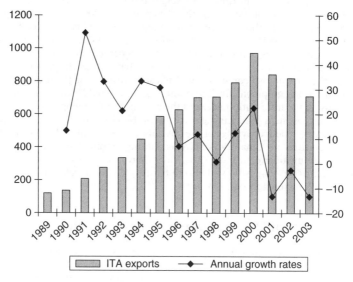

Figure A.1   Trend in world export of ITA goods ($ billion)
Source: Based on Bora (2004).

trends continued since ITA till 2000, though with a lower growth rate (11.5 per cent), the period thereafter showed steady decline in the world exports. Not only that the rate of growth could not be sustained since ITA, but also the observed growth rates since 2000 have been negative (−9.9 per cent during 2000–2003). As a result, the level of world exports of ITA goods in 2003 ($710 billion) turned out to be almost at the level that prevailed in 1997 ($701 billion). Thus viewed, the trade liberalization as envisaged under ITA had only limited impact in promoting world demand for IT goods through increased competition and reduced prices.

Members vs non-members of ITA

Let us examine how has different groups of countries identified above performed with respect to trade in ITA goods. ITA goods may be divided broadly into six different categories, namely computers, telecommunication equipment, semiconductor manufacturing equipment, instruments, components and software traded through electronic media. According to Bora (2004) in 2002 Computers accounted for about 35 per cent of the trade in IT goods followed by semiconductor manufacturing equipment (30 per cent), telecom equipment (19 per cent), components and parts (12 per cent) and instruments (3 per cent). Software (traded through electronic media) accounted for only about 1 per cent.

For our analysis we have selected three products – computers, telecom equipment and components – that accounted for about 66 per cent of the total trade in ICT goods in 2002. The recorded rate of growth of exports and imports of these products by the three different groups of countries identified and for the world is presented in Table A.1. It is evident that the developing ITA members as a group performed better as compared to the developed ITA members in terms of exports and imports of all the three major product groups considered here. The only exception has been in the exports of components where the average growth rate recorded by the developed ITA members is higher than the developing ITA members. But when we compare the performance of developing ITA members with the developing non-ITA countries there is not much room for cheer. In

Table A.1 Annual average growth rate in exports and imports of select ITA products by different group of countries (1999–2003)

Products	Exports/ imports	Developed ITA members	Developing ITA members	Developing non-ITA members	Total
Computers	Imports	1.80	15.60	24.99	3.71
	Exports	−3.85	14.61	35.13	5.14
Telecom	Imports	6.25	13.75	13.04	7.35
	Exports	−0.31	22.95	28.99	6.75
Components	Imports	3.99	9.20	28.15	8.78
	Exports	5.04	4.00	21.97	4.69

Source: International Trade Centre, PC-TAS CD-Rom Database.

the entire ITA product groups under consideration, the exports and imports by developing non-ITA member countries is found to be much higher than their counterparts who joined ITA.

ICT diffusion in developing countries

The ITU estimates show that during 1992–2002 the share of developing countries in fixed telephones increased from 21 to 45 per cent, mobiles from 12 to 46 per cent, PC users from 10 to 27 per cent and most impressively Internet users from 3 to 34 per cent (UN ICT Taskforce 2005). Though these achievements are by no means less impressive, if we look at the use of ICT in terms of the number of telephone lines in the developing countries of different regions we find significant inter-regional variation. As is evident from Table A.2, the telephone density in Sub-Saharan Africa, Pacific and South Asia is only one-tenth of that in Caribbean. Thus observed growth has been concentrated in few regions. More importantly, it appears that the observed growth has been mainly on account of the pent-up demand in urban areas of developing countries leading to significant intra-national digital divide. The point needs to be illustrated. In 1996, the teledensity (fixed) in South Asia and in the largest city has been 1.4 and 6.9 respectively. By 2002, while the teledensity in the largest city of South Asia increased to reach 12.7, that for the region as a whole remained at 3.4. We have also seen that in countries like Lao PDR, Myanmar and Cambodia the capital city alone accounted for more than 70 per cent of the telephone lines in these countries.

In terms of the NRI (Dutta *et al.* 2003), Brazil with a rank of 29 was positioned above countries like Malaysia, India, Thailand and others who are members of ITA. Similarly Chile and South Africa were ranked above India and Thailand. Similar observations could be made in terms of other indicators like household spending on IT, telephone intensity and other indicators of IT use. Thus viewed, trade liberalization as undertaken under ITA seems to have not enabled the developing ITA members to overcome the limits set by their per capita income to promote IT use.

Table A.2 Number of telephones (fixed and mobile) per 100 inhabitants in the developing countries of different regions in 1992 and 2002

Regions	1992	2002
Caribbean	9.88	52.6
Europe and Central Asia	14.1	44.1
Latin America	7.1	35.4
Middle east and North Africa	4.5	18.0
East Asia	1.2	27.4
Sub-Saharan Africa	1.0	5.3
Pacific	2.3	4.7
South Asia	0.7	4.5

The empirical evidence, though preliminary, tends to suggest that trade liberalization, as envisaged under ITA, *per se* may not be adequate to enabling the developing countries to enjoy the fruits of ICT revolution. Hence much could be gained by the developing ITA members, and ITA would become more attractive to other developing countries if the trade liberalization is complimented with targeted efforts at capacity building which is currently missing in ITA.

Notes

1 Introduction

1. For a detailed discussion on General Purpose Technology, see Bresnahan and Trajtenberg (1995).
2. The productivity paradox has been best summarized by referring to Solow's (1987) much quoted statement – 'you can see computer age everywhere, but in the productivity statistics'.
3. See Appendix 1 for more details on ITA.
4. In this study, the terms IT and ICT has been used interchangeably.
5. For more details on India's credentials in the ICT software and service sector, see Chapter 2.
6. In this book these four countries are denoted as CLMV countries or ASEAN New Comers.
7. See in this context Wong and Singh (2002).
8. The report argues that actual inflation fell by 0.5 per cent points a year from 1994 to 1998 due to the effect of declining prices of IT goods. Also IT industry, including telecommunications, employed 7.4 million workers in 1998 and this accounted for 6.1 per cent of the total employment with an annual wage rate more than 1.5 times that for all private employees.
9. See Joseph (1997) for a detailed analysis of the product structure of electronics industry and the implication of product structure for the growth performance.
10. This has been attributed mainly to the fact that labor is the major input in the production of services, and the abundant supply of labor in less developed countries translate into low wages. Since the technology of producing services does not differ significantly across countries, lower wages results in low cost of production of services in less developed countries (Bhagwati 1984).
11. For a recent review of studies the readers are referred to Indjikian and Siegel (2005).
12. See Parthasarathi and Joseph (2004) for a discussion about the use of ICT in other sectors like oil and natural gas, railways and others.
13. Here it may be noted that the empirical evidence pertains mainly to the developed countries. When it comes to the less developed countries, though there are a number of cases where ICT has been used for addressing the development issues, most of them appear to be in the pilot stage (see for details Joseph 2002, Carlos et al. 2003).
14. Here the readers are referred to the large number of papers presented in the GLOBELICS conferences held in Rio in 2003, Beijing in 2004 and Pretoria in 2004. The papers are available at www.globelics.com.

15. While the former is the capability of operating or imitating the production of a particular machine, the latter involves knowledge of the underlying scientific and design principles.
16. While the NSI framework highlights the importance of such interactions promoting learning and innovations, this remains a relatively unexplored area in most developing countries.
17. The list of MNEs that have set up global R&D centers in India includes Akzo Nobel, AVL, Bell Labs, Colgate Palmolive, Cummins, Dupont, Daimler-Chrysler, Eli Lilly, GM, HP, Honeywell, Intel, McDonald's, Monsanto, Pfizer, Texas Instruments, Unilever, among many others.

2 India: An IT powerhouse of the South

1. Limits on space prevented from venturing into a detailed analysis of the growth performance of agriculture, industry and service sectors. Interested readers are referred to, among many others, Basu (2003), Nagaraj (1997, 2003), Gulati (1998), Bhide and Kalirajan (2003).
2. Bangalore, a single location in India, alone has over 140,000 IT professionals, 20,000 more than Silicon Valley!
3. As quoted in *Economic Times*, 21 July 2004.
4. It has been estimated that on the average the labor cost in India in the ITES sector is only about 14 per cent of that in US.
5. For detailed discussion on Software Product Development experience in India, see Krishnan and Prabhu (2004).
6. See for details http://www.bangaloreit.com/html/itsckar/itindustriesother-cities.htm.
7. This refers to the NIC codes 35 and 36 at the two digit level.
8. More than 800 e-governance projects have been initiated either by the state or by the central government. These projects are at different stages of their implementation. See for details: www.egovdatabase.gov.in.
9. See Singh (2002) and Kaushik and Singh (2004) for a detailed discussion of ICT initiatives for rural development at the instance of TARAhaat and Drishtee in states of Haryana and Punjab.
10. For an interesting critique of different IT initiatives for development, see Sreekumar (2002).
11. See in this context India, Electronics Commission (1971) Proceedings of the National Conference on Electronics, Bombay, 24–28 March (Bombay: Electronics Commission).
12. Mention need to be made of the substantial reduction in the duties and tariffs across the board for components and sub assemblies, zero duty of software import and zero income tax on profits from software exports.
13. For a detailed comparative analysis of the policies initiated by different state governments against the backdrop of the national policies, see Joseph (2004a). Heitzman (1999) provides a detailed analysis of the role of the Government of Karnataka in making Bangalore the Silicon Valley of India.

14. A STP in all respects is similar to a free trade zone exclusively for the software. The specific objectives of the STPs are:

    - to establish and manage the infrastructural resources such as data communication facilities, core computer facilities, built-up space, common amenities and so on;
    - to provide services (import certification, software valuation, project approvals, etc.) to the users who undertake software development for export purposes;
    - to promote development and export of software and software services through technology assessments, market analysis, marketing support and so on;
    - to train professionals and to encourage design and development in the field of software technology and software engineering (Joseph 1997).

15. "STPI now opens office at Silicon Valley, USA", *The Economic Times*, New Delhi, Special Supplement on Software Technology Parks of India, June 11, 2000.

16. For a detailed discussion on telecom policies and performance, see ITU (2002a).

17. To begin with, there was the Computer Society of India, which is essentially an association of academics and professionals and did not address many of the issues faced by the industry. Hence a new association called Manufacturers Association of Information Technology (MAIT) was formed in 1982. This consisted of both the hardware and the software firms. Later an association, currently known as Nasscom, was formed to address specific issues being faced by the software and service companies. The Electronics and Software Export Promotion Council, an autonomous body under the MIT, through its various initiatives also made significant contribution towards India's IT export growth.

18. For a detailed account of the NASSCOM activities in promoting IT and role played by late Dewang Metha, see "Power Lobbying", *Business India*, February 19 to March 4, 2001.

19. The aforesaid royalty limits are net of taxes and are calculated according to standard conditions.

3 Thailand: From investment-led growth to innovation-led growth

1. Fujitsu, a subsidiary of Fujitsu of Japan, completed a $425 million expansion project in 2001 to increase annual production of hard disk drives to 29.1 million units. Fujitsu employs 5000 workers at its two manufacturing plants in Thailand.

2. Seagate invested $115 million in 2003 to improve its production process and upgrade labor skills at three manufacturing plants. Seagate has its largest production base in Thailand with 33,000 employees and 5 manufacturing plants.

3. In 2003 Lucent Technologies Microelectronics, owned by Lucent of Murray Hill, New Jersey, completed a $30 million expansion project to double its production of ICs. The plant has 1200 employees and is one of its largest IC assembly and test operations worldwide.

4. AMD (Advance Micro Devices), headquartered in Sunnyvale, California, also has expansion plans in Thailand and has 1300 employees.

5. NS Electronics made an additional investment of $40 million in 2003 on machinery and equipment to produce new product packages. The firm also plans to increase its production capacity of ICs from 2.5 to 5 million units per day. NS Electronics is a Thai company, separated from Alphatec Group and maintains strong position as an export manufacturer. The firm has 2800 employees.

6. For a critical review, refer Thajchayapong (1997).

7. For a discussion, see Nongphanga Chitrakorn (1990).

8. This refers to new software technology, such as Java and NET, used for building software systems that operate on network.

9. For a detailed analysis of various policies on Human Development and for the present status, see S. Suksiriserekul (2003) and Lauridsen (2002).

10. Draws from Board of Investment (2003), *A Guide to the Board of Investment*, Bangkok.

11. TOT operates in the metropolitan and provincial areas. Also, it provides international service to neighboring countries. Telecom Asia operates in the Metropolitan areas, TT&T in the provincial area and CAT had the monopoly of international telecom services. See in this context Asia Info-communication Council Member Information Thailand, www.aic.or.jp/relatedcountryper centstatus20per centreport.thailand.htm.

12. The Telecom Asia's contract with TOT covered the provision of 2 million lines (later extended to 2.6 million) in the Bangkok region. Telecom Asia is founded around the Charoen Pokphand (CP) Group, which is Thailand's largest conglomerate through its agricultural and food processing activities. It owned 22 per cent of Telecom Asia at end 2000. Telecom Asia's foreign partner is NYNEX of the US, which owns 18 per cent. Part of Telecom Asia was later listed in the Thai stock market. See for details, ITU (2002).

13. The TT&T www.ttt.co.th contract covered the provision of 1 million lines (later extended to 1.5 million) in the provinces. TT&T has a number of owners with shareholders that include Jasmine International (20 per cent), Loxley (13 per cent), Italian-Thai Development (10 per cent), Thai Farmers Bank and NTT West (18 per cent) of Japan. By June 2001, TT&T had activated some 1.19 million lines but it is growing at a slow rate (less than 2 per cent per year). Both Jasmine and Loxley have their own ISPs.

14. In 2000 it was argued that Thailand installed some 7.4 million fixed telephone lines, which equates to just over 8.5 telephone lines per hundred people. Of these, 4.5 million are in the Bangkok area, serving 15 per cent of Thailand's population. The remaining 2.9 million are installed in the provinces and serve more than 50 million people. http://www.andrew.com/access/0801/articles/thailand.asp.

15. The waiting time for a telephone in Thailand is said to have grown to almost eight years by the middle of 1980s.

16. Build–Transfer–Operate Scheme is something unique to Southeast Asia and has its roots in Thailand. As per the scheme, a few firms called "concessionaries" are given the permission to invest in the area for a specified period of time on a revenue-sharing basis, after which the assets made are to be transferred to the state. Viewed thus, the firms do not own the assets that they generate.
17. See ITU (2003) for more details on the terms of contract.
18. http://www.school.net.th.
19. See for a discussion, Pongsrihadulchai (1998).
20. See for details, NECTC, 2002, *ICT for Poverty Reduction: Examples of Programmes/Projects in Thailand*, NECTEC, Bangkok.

4 Cambodia: Between Pentium and Pencillin

1. For an excellent treatment of the various issues in the structural adjustment process of Cambodia in a historical perspective, see Kannan (1997).
2. Available data shows that the total railway length in the country remained at 601 km during the last two decades.
3. See for details www.nida.gov.kh/activities/summit.
4. See for details http://www.nida.gov.kh/activities/ict_policy/ictdraft.pdf.
5. It should also be noted that that only about 9 per cent of the households in Cambodia are having access to electricity.
6. For a pioneering and excellent document on Cambodia's telecom and Internet, see ITU (2002).
7. Camintel works with a 128 kbps line leased from Camnet. Camintel plans to become a fully independent ISP with its own network and infrastructure.
8. In practice, investors report that licensing requirements involve significant red tape and visits to multiple government agencies, despite CDC's designation as a one-stop shop.
9. For an excellent treatment of Trade policy issues, see Center for International Economics (2001).
10. The Working Groups are Export Processing Industries; Manufacturing and Distribution; Energy and Infrastructure; Tourism; Law, Tax and Governance; Banking, Financial Services and Insurance; and Agro-processing.

5 Lao PDR: Hastening slowly?

1. The small-scale units are defined as those employing less than 10 persons and medium refers to those employing 10–99. The large-scale units are the ones employing more than 99 persons.
2. It was in 1990 that the compilation of GDP in Lao PDR started using the value-added approach.
3. These estimates are based on the data obtained from National Statistical Center (2000, 2002).
4. Since the beginning of economic reforms in 1988, approximately 90 per cent of state-owned enterprises have been converted to another system of

management (many via leases) or liquidated (US Department of Commerce 2002b).
5. This project is being undertaken with the assistance of Francophone Organization.
6. See for details http://seamedia.org/Lao PDR.
7. Based on Xaphakdy Smith, Developing Information and Communication Infrastructure: An Implementation of Tokyo Declaration and Action Plan in Lao PDR, http://unpan1.un.org/intradoc/groups/public/documents/apcity/unpan006169.pdf.
8. For a detailed account of higher education in Lao PDR, see John C. Weidman (undated) Reform of Higher education in the Lao People's Democratic Republic, Paper presented at the 1995 Annual Meeting of the Association for the Study of Higher Education, Orlando, Florida, 1 November. Revised version published under the title, "Lao PDR" (Chapter 5), in Gerard A. Postiglione and Grace C.L. Mak (eds), *Asian Higher Education: An International Handbook and Reference Guide*. Westport, CT, USA: Greenwood Press, 1997.
9. For a detailed account of the IT education in Lao PDR, see UNDP (2002).
10. The UNDP study reported that the FEA intends to offer a course on Computer Engineering Program in the near future.
11. As part of the bilateral agreement with the Government of India, 45 government officers underwent computer training in India during 2002.
12. From the Web site of Jhai foundation it is understood that it provides ICT access in a secondary school at Phonsong and the children are found highly enthusiastic about learning computers. If the experience of Jhai foundation is any indication there appears an urgent need for replicating these models at the instance of other NGOs operating in the country.
13. These observations are to be viewed with sufficient caution, for the "cause and effect" relationship is not statistically established on account of the lack of needed data.
14. During the discussion with concerned officers and business community, it was informed that this anomaly has been corrected recently. This again indicates the proactive nature of the government and its commitment to do away with the hurdles.
15. For details visit http://www.ediindia.org.
16. Here it needs to be noted that this issue has already been recognized by the multilateral agencies like UNESCAP and it has already undertaken a series of training programs towards addressing this issue. See GMS forum (2002) Corporate Strategy/Business Plan for the GMS Business forum, paper jointly developed by UNESCAP, ADB and GMS Business Forum Secretariat and was adopted at the GMS Business Forum Third Annual Meeting, Kunming, 12–13 December.
17. Going by the information provided by the country commercial guide published by the US Department of commerce (2002b), in 1998, the total market for ICT hardware and accessories accounted for $4.17 million and it was met entirely through imports. Similarly the total electronics demand in the country was estimated at $24.88 which was also met entirely by imports. For details see US Department of Commerce (2002b). http://usembassy.state.gov/Lao PDR /wwwhcotg.html.

18. Three more companies are in the process of obtaining ISP license from LANIC (Phissamay 2002).
19. The actual number of Internet users in the country is bound to be much larger. Estimates indicates that as of September 2001 there are around 9000 users (Phissamay 2002, ITU 2003).
20. See for details, Carlos A. Primo Braga, John A. Dally and Bimal Sareen (2003), "The future of Information and Communication Technology for Development", paper presented in the ICT Development Forum, Petersberg, Germany.
21. Jhai Foundation is a reconciling organization founded by an American air war veteran, Lee Thorn, and a refugee from the bombing. Jhai foundation work with friends in Lao PDR to create what they want: schools, computer labs, and weaving and coffee cooperatives. See for details http://www.jhai.org.
22. Here the reference is to the CATNET service wherein public Internet Kiosks have been set up with a view to narrow the digital gap at the instance of Communication Authority of Thailand (CAT). See for details Rattakul Rattananwan (2003), "Bridging the Digital Divide: A Case Study of CATNET Nationwide Internet Kiosks", in Proceedings: The Regional Conference on Digital GMS, Asian Institute of Technology, Bangkok.
23. The Simputer is a low-cost portable alternative to PCs by which the benefits of IT can reach the common man. It has a special role in the third world because it ensures that illiteracy is no longer a barrier to handling a computer. The key to bridging the digital divide is to have shared devices that permit truly simple and natural user interfaces based on sight, touch and audio. The Simputer meets these demands through a browser for the Information Markup Language (IML). IML has been created to provide a uniform experience to users and to allow rapid development of solutions on any platform. See for details http://www.simputer.org.

6 Myanmar: Sowing but not harvesting?

1. We have been compelled to use the World Bank data because the data provided by the Central Statistical Organization (CSO) follows a classification different from those adopted by most countries. Instead of following the conventional division of GDP into primary, secondary and tertiary sectors, the CSO divides the GDP into following three categories, namely goods, services and trade.
2. These objectives are published in an English Newspaper called *The New Light of Myanmar Published by the Government*.
3. All the estimates presented above are based on the data provided by CSO (2001).
4. Please note that the figures are in current prices at the time of writing and hence are overestimates.
5. The Computer Science Development Council is a statutory body having the constitutional powers and responsibilities which include, among other things, formulation of appropriate policies for the growth of IT sector and for promoting the use of IT in the country.
6. See for details, http://www.myanmar.com.gov/laws/computerlaw.html.

7. The CSDC has a person assigned by the State Law and Order Restoration Council as the chairman and the following as the members: minister or deputy ministers from relevant ministries, heads of the relevant government departments and organizations, suitable computer scientists, and the deputy minister of education as the secretary.

8. The data is according to the 1983 Census as reported in CSO (2001).

9. CSO (2001).

10. Myanmar has not purchased its own communication satellite, but it leases transponders aboard several regional satellites, such as AsiaSats and the Indonesian Palapa_C satellite. Myanma Post and Telecommunications also has an agreement to use ThaiCom-3 satellite transponders in its internal SATCOM services.

11. See for details http://www.myanmars.net/myanmar/internet.htm.

12. See for details "Bagan Cybertech offers New Services in Mandalay", *The Myanmar Times*, April 28–May 4, 2003.

13. This may be compared with a monthly fee of about $650 in India without any limit for data transfer. See http://www.ddsl.net/.

14. See for details "Senior officials go back to school over e-government", *Myanmar Times*, June 30–July 6, 2003.

15. Priority item A includes agricultural machinery, and farm implements, fertilizers, pesticides, high-yield quality seeds edible oil and fats for soap industry, construction stores and building material.

    Priority Item B includes about 60 items grouped under personal goods, household goods, food stuff, construction materials, textile products, electric and electronic products, and general products.

16. Yet, it was learned from discussions with the senior officials at Myama Post and Telecom that there are about 40,000 mobile subscribers and most of them are in the urban areas.

17. see for details http://www.abroad-phone-rental.com/Myanmar_iridium_satellite_phone_rental.htm.

18. See for details, "Billions of Kyats, Millions of Dollars spent in developing ICT infrastructure and facilities", Opening address given by Secretary – I at the Second Annual Myanmar ICT week.

7 Vietnam: Another tiger in the making?

1. In 1990, the share of primary sector and that of tertiary sector in GDP was almost equal at a little more than 38 per cent and that of secondary sector was 22.7 per cent. By 2002, the share of agricultural sector recorded a decline of more than 15 per cent point to reach a level of 23 per cent. The decline in the share of primary sector was almost entirely compensated by the secondary sector with an increase in its share from 22.6 per cent in 1990 to 38.5 per cent in 2002. Changes have also been taking place, *albeit* slowly, in the ownership structure in the economy. The most spectacular change has been an increase in the share of foreign investment sector in GDP from 6.3 per cent in 1995 to over 13 per cent in 2001.

2. This is not to ignore the decline in the rate of industrial growth from two-digit level in almost all the years of 1990s to 8 per cent in 1997–98 and

7 per cent in 1998–99. The recorded growth rate of the service sector was only 2.5 per cent during 1998–99 and that of agricultural sector almost halved from 8 to 9 per cent in most of the years up to 1996–97. Estimates are based on data obtained from General Statistical Office (2001).

3. Majority of the approximately 6000 state-owned enterprises were barely breakeven and the rate of overemployment in 1996 was estimated at 25 per cent (Kokko 1998).

4. The BTA is a comprehensive agreement that covers trade in goods, services, investment, intellectual property, the right of appeal and transparency matters, as well as general business facilitation issues. As per BTA, the US has granted Vietnam "normal trade relations" status (which implies the application of the MFN principle, MFN), removing Vietnam from a small list of countries which were denied this grant, including North Korea, Afghanistan, Serbia and Cuba. See for details, the statement by Jeffrey A. Bader, Deputy Assistant Secretary for East Asian and Pacific Affairs, to the House Committee on International Relations, Washington DC, June 18, 1997, available at http://usinfo.state.gov/regional/ea/vietnam/bader618.htm.

5. Some countries still enjoy duty advantages in the US as compared to Vietnam, such as Mexico and Canada (under NAFTA), Jordan (under its Free Trade Agreement) and certain Caribbean nations (under the Caribbean Basin Initiative).

6. See for details Bezanson *et al.* (1999).

7. Based on Directive No. 58-CT/TW on accelerating the use and development of information technologies for the cause of industrialization and modernization, dated October 17, 2000.

8. For more details see, Decision No. 81/2001/QD-TTG of the prime minister on approving the Action Plan to implement the Directive No. 58-CT/TW of the communist party of Vietnam, dated May 24, 2001.

9. See for details; Decision No. 128/2000/QD-TTG of the prime minister on a number of policies and measures to stimulate investment and development of software industry. (For more details see the next section.)

10. See for details; Circular No. 31/2001/TT-BTC of the Ministry of Finance on the instructions of the implementation of preferential tax treatment regulated at Decision No. 128/2000/QD-TTG of the prime minister.

11. Sudhir Kumar, Foreign Presence in the field of IT in Vietnam, Indian Embassy in Vietnam (mimeo), undated.

12. The Melbourne-based RMIT has announced a US$30 million expansion plan. RMIT, which currently has about 700 students at both graduate and postgraduate levels (in IT and other fields) in its facilities in HCMC and other places has a target to reach 10,000 students at its new educational center in HCMC.

13. According to VNPT, the postal side of its business employs 49 per cent of the organization's staff, yet generates only 5–7 per cent of gross revenue. "VNPT to be first economic group in Vietnam?", June 3, 2001, See in this context http://db.vnpt.com.vn/News/viewutf8.asp?ID=2740.

14. See for details the Web site of the Communist Party of Vietnam, www.cpv.org.vn/chuyende/nationalcongress9/docs/jan02_vietnampost.htm.

15. Internet in the country had its origin in 1991 when the possibility of exchanging e-mail with a German University was explored but proved to be

unfeasible. In 1992 Institute of Information Technology, Hanoi, established a dial-up connection with the Australian National University to exchange e-mail. For a brief history, see ITU (2002).

16. For a succinct historical background of foreign investment in Indo-China, see Freeman (2002).

17. Here the approval is given for a foreign firm for construction and operation of an infrastructure project within a certain period of time; on the expiry of the period the foreign investor shall transfer without indemnification, the project to the state of Vietnam.

18. Here, again the contract is for the construction of an infrastructure project; on the completion of the construction the foreign investor shall transfer the project to the state. The government shall give the investor the right to operate the project within a certain period of time so as to retrieve invested capital and earn reasonable profit.

19. In this case the foreign investor, on the completion of the project, shall transfer the project to the state. The Government of Vietnam shall create conditions for the foreign investor to carry out another project so as to retrieve the invested capital and earn reasonable profits.

20. See for details Decision No. 128/2000/QD-TTG of the prime minister on a number of policies and measures to stimulate investment and development of the software industry.

21. However, going by Freeman's estimate the stock of per capita FDI in Vietnam in 2000 ($225) is found to be much lower than Malaysia ($234) or Thailand ($388).

22. For China and India the period covered refers to 1991–99. See for details, Joseph (2002).

23. There are a large number of studies dealing with the trade and investment liberalization in Vietnam. See in this context, among others, Center for International Economics (1999), Duong (undated), US Department of Commerce (2002c) and UNESCAP (2001).

24. Based on discussions with government officials.

25. Quoted from www.emich.edu/ict_usa/Vietnam.htm.

26. For an excellent analysis of the state of telecommunications, Internet and other related aspects in Vietnam, see ITU (2002).

27. Mobiphone is being operated under the BCC revenue-sharing and technology-sharing arrangement between VNPT and Millicom, which is based in Luxumburg, having 90 per cent share holding in the Swedish company Comvik. The local company is called Comvik International Vietnam.

28. See for details Ha Thuy, Vietnam Internet: Challenges and Prospects, *Vietnam Economic Times*, No. 14. 31/5/2001.

29. It needs to be noted that the firms in the STPs are eligible for a discount of 50 per cent. Yet the leased-line cost is high by international standards and also in comparison with countries like India. Cost of 2 Mbps line in India was approximately $650 without any limit to data transfer and the cost for firms in the STPs is still lower. see http://www.ddsl.net/.

30. Draws from the Decision No. 11/2001 dated 25 July, of the prime minister on ratifying the project on the state administrative management computerization in the 2001–2005 period.

31. CAPNET, established under the prime minster's Decision No. 280/TTG, dated April 29, 1997, has been designed and built up according to the divisible architecture of the state administrative bodies and divided into the following levels.

> Level A: The government
> Level B: The ministries, provinces
> Level C: The provincial/municipal services, departments, branches districts, provincial towns or departments and units attached to the ministries
> Level D: Communes and wards.

8   ICT and the developing countries: Towards a way forward

1. It is a low-cost fixed wireless access technology aimed at connecting primarily homes and small offices in rural areas and small towns. CorDECT provides two lines to each subscriber, a voice line and a 35 kbps dedicated Always-ON Internet connection (a premium rate at 70 kbps). Capable of being used in both rural and urban areas, its cost-effectiveness is highlighted better in the rural case where using the Relay Base stations it can serve users in a radius of 25–30 km. Such rural deployment costs less than $300 per line, making CorDECT the lowest cost connectivity solution. See for details Best, M.L., Wither Wireless Networks for Rural Development, www.media.mit.edu/~mikeb/**wither.pdf** and CorDECT occupies pride of place, *The Hindu*: http://www.thhindu.com/thehindu/seta/2003/03/06/stories/2003030600160200.htm.
2. Jhai Foundation is an organization founded by an American air war veteran, Lee Thorn, and a refugee from the bombing. Jhai foundation works with schools, computer labs, and weaving and coffee cooperatives. See for details http://www.jhai.org.
3. Simputer (simple computer) was developed by scientists from the Indian Institute of Science in Bangalore and Encore, a software company. At US$200 each (in 2002), Simputer offers computing facilities at a drastically lower cost compared to US$650 for a PC. Apart from the low cost, simputer has many other advantages: it is roughly the size of a hand-held electronic organizer, thereby making it really portable. It can run on AAA battery, thereby not relying on power connection, which in rural India is unreliable. It uses IML (Information Mark-up Language) to convert English content (from the Internet) into many local languages. It has a text-to-speech converter that reads out the content. See for details Sukumar and Narayanamurthy (2003).

Appendix   ITA: Addressing digital divide through trade liberalization

1. Studies have shown that the inter-country differences in rate of IT diffusion are significantly related to general levels of socio-economic development represented by per capita GDP, R&D expenditure and the levels of human development (Hargittai 1999, Rodriguez and Wilson 2000, Norris 2001).

2. The exact text of the ITA, including the product coverage, can be found at http://www.wto.org/english/docs_e/legal_e/itadec_e.htm.
3. This refers to the list of 16 developing countries that signed ITA by 1997.
4. We have selected the following ten non-ITA developing countries: China, Chile, Mexico, Egypt, Brazil, Argentina, Hungary, Russia, South Africa, Iran (Islm.R). China has been included in this group as it joined ITA only in 2002.

Bibliography

ADB (2004) *Asian Development Outlook 2004* (Hong Kong: Oxford University Press).

Alcorta, L. and W. Peres (1998) "Innovation Systems and Technological Specialization in Latin America and the Caribbean", *Research Policy*, 26(7–8): 857–881.

Arnold, E., M. Bell, J. Bessant and P. Brimble (2000) *Enhancing Policy and Institutional Support for Industrial Technology Development in Thailand: The Overall policy Framework and Development of the Industrial Innovation System* (Bangkok: Techno polis).

Arocena, R. and J. Sutz (2000) "Looking at National Innovation Systems from the South", *Industry and Innovation*, 7(1): 55–75.

Arora, A. and S. Athreya (2002) "The Software Industry and India"s Economic Development", *Information Economics and Policy*, 14(2): 252–273.

Arora, A. and A. Gambardella (2004) "The Globalization of the Software Industry: Perspectives and Opportunities for Developed and Developing Countries", Working Paper No. 10538, National Bureau of Economic Research, Cambridge, MA.

Arora, A., V. S. Arunachalam, J. Asundi and F. Ronald (2001) "The Indian Software Services Industry", *Research Policy*, 30(8): 1267–1287.

Bank of Thailand (2003) *Economic and Financial Statistics, First Quarter 2003* (Bangkok: Bank of Thailand).

Basu, Kaushik (2003) "The Indian Economy: Up to 1991 and Since", BREAD Working Paper No. 052, available at http://www.cid.harvard.edu/bread/papers/working/052.pdf.

Bell, M. and K. Pavitt (1992) "*Accumulating Technological Capability in Developing Countries*", Proceedings of the World Bank Annual Conference on Development Economics, pp. 257–281.

——(1997) "Technological Accumulation and Industrial Growth: Contrasts Between Developed and Developing Countries", in D. Archibugi and J. Michie (eds), *Technology, Globalisation and Economic Performance* (Cambridge: Cambridge University Press), pp. 83–137.

Bell, M., Scott Kemmis and W. Satyarakwit (1982) "Limited Learning in Infant Industry: A Case Study", in F. Stewart and J. James (eds), *The Economics of New Technology in Developing Countries* (London: Frances Pinter).

Bezanson, K. J., K. Annerstedt, D. Chung, G. Hopper, Oldham and F. Sagasti (1999) *Vietnam at the Crossroads: The Role of Science and Technology* (Canada: IDRC).

Bhagwati, J. N. (1984) "Why are Services Cheaper in Poor Countries?", *Economic Journal*, 94: 279–285.

Bhide, S. and K. P. Kalirajan (2003) "Impact of Sector-specific and Macro-level Reforms on Agriculture: Simulations of a Macroeconometric Model", in K. P. Kalirajan and U. Sankar (eds), *Economic Reform and the Liberalization of the Indian Economy: Essays in Honour of Richard T. Shand* (Cheltenham: Edward Elgar).

Binh, N. N. and J. Houghton (2002) "Trade Liberalization and Foreign Direct Investment in Vietnam", *ASEAN Bulletin*, 19(3): 302–318.

Board of Investment (2003) *A Guide to the Board of Investment* (Bangkok: BOI).

Bora, B. (2004) "Information Technology Agreement and World Trade", Presentation made at the Information Technology Symposium, organised by WTO, Geneva, October 18–19.

Bresnahan, T. and M. Trajtenberg (1995) "General Purpose Technologies: Engines of Growth", *Journal of Econometrics*, 65(1): 83–108.

Breschi, S. and F. Malerba (1997) "Sectoral Innovation Systems: Technological Regimes, Schumpeterian Dynamics and Spatial Boundaries", in C. Edquist (ed.), *Systems of Innovation: Technologies Institutions and Organizations* (Washington, London: Pinter).

Brooker Group (2001) *Technical Innovation of Industrial Enterprises in Thailand 2001* (Bangkok: Brooker Group).

Ca, T. N. (1998) "Technology Assessment in Vietnam: Concept and Practices", *Science and Public Policy*, 25(2): 87–94.

Cairns, R. D. and D. Nikomborirak (1997) *An Assessment of Thailand"s New Telecommunications Plan* (Bangkok: Thailand Development Research Institute).

Cantwell, J. A. (1995) "The Globalization of Technology: What Remains of the Product Cycle Model?", *Cambridge Journal of Economics*, 19(1): 155–174.

Carlos, A., P. Braga, J. A. Dally and B. Sareen (2003) "The Future of Information and Communication Technology for Development", Paper presented in the ICT Development Forum, Petersberg, Germany.

Carlsson, B. and R. Stankiewiez (1995) "On the Nature, Function and Composition of Technological Systems", in B. Carlsson (ed.), *Technological Systems and Economic Performance: The Case of Factory Automation* (London, Boston, Dordrecht: Kluwer Academic Publishers).

Carlsson, B., S. Jacobsson, M. Holmen and A. Rickne (2002) "Innovation Systems: Analytical and Methodological Issues", *Research Policy*, 31: 233–245.

Cellular Operators Association of India (2005) COAI Annual Report 2005, available at http://coai.in/DOCS/COAI%20Annual%20Report%202005.pdf.

Center for International Economics (1999) "Trade and Industry Policies for Economic Integration", Report prepared for CIEM and UNIDO, Centre for International Economics, Canberra, http://www.thecie.com.au/pdf/ciem.pdf.

——(2001) *Component Reports: Macro Assessment, Trade Policy, Trade Facilitation, Poverty Assessment – Part B* (Phnom Penh: Ministry of Commerce).

Chidamber, S. R. (2003) "An Analysis of Vietnam"s ICT and Software Services Sector", *The Electronic Journal on Information Systems in Developing Countries*, 13(9): 1–11, http://www.is.cityu.edu.hk/research/ejisdc/vol13/v13r9.pdf.

Choudhary, B. (2001) "Role of Foreign Direct Investment in the Chinese Economy with Special Reference to the Overseas Chinese: Implication for India", *China Report*, 37(4): 463–474.

Chu Huu Quy (1996) *All Sided Economic Development of Rural Vietnam* (Hanoi: National Politics Publishing House), p. 18.

Cooper, C. (1980) "Policy Interventions for Technological Innovations in Developing Countries", Staff Working Paper No. 441, The World Bank, Washington, DC.

CSO (2001) *Statistical Yearbook 2001* (Yangon: Central Statistical Organization).

Dahlman, C. J. (1984) "Foreign Technology and Indigenous Technological Capability in Brazil", in Martin Fransman and K. King (eds), *Technological Capability in the Third World* (London: Macmillan), pp. 317–334.

Dahlman, C. J. and P. Brimble (1990) "Technology Strategy and Policy for Industrial Competitiveness: A Case Study of Thailand", Industry and Energy Department Working Paper (Industry Series) No. 24, World Bank, Washington, DC.

Dahlman, C. and L. Westpal (1982) "Technological Effort in Industrial Development: An Interpretive Survey", in F. Stewart, and J. James (eds), *The Economics of New Technology in Developing Countries* (London: Frances Pinter).

David, P. A. (1990) *The Dynamo and the Computer: An Historical Perspective on the Modern Productivity Paradox*, AEA Papers and Proceedings, 355–361.

——(1991) "Computer and Dynamo: The Modern Productivity Paradox in a Not-Too-Distant Mirror", OECD, *Technology and Productivity: The Challenge for Economic Policy* (Paris: OECD), pp. 315–348.

D"Costa, A. P. (2003) "Uneven and Combined Development: Understanding India"s Software Exports", *World Development*, 31(1): 211–226.

——(2004) "Export Growth and Path-Dependence: The Locking-in of Innovations in the Software Industry", in A. P. D"Costa and E. Sridharan (eds), *India in the Global Software Industry: Innovation, Firm Strategies and Development* (New York: Palgrave Macmillan).

Deasi, A. V. (1984) "India"s Technological Capability: An Analysis of its Achievements and Limits", *Research Policy*, 13: 303–310.

DeBresson, C. (1989) "Breeding Innovation Clusters: A Source of Dynamic Development", *World Development*, 17(1): 1–6.

DeBresson, C. and F. Amesse (1991) "Network of Innovators: A Review and Introduction to the Issue", *Research Policy*, 20: 363–379.

Department of Skill Development (2000) *Direction in skill Development, Technical Studies and Planning Division* (Bangkok: Department of Skill Development, Ministry of Labour and Social Welfare).

DOI (2001) *Creating a Development Dynamic: Final Report of the Digital Opportunity Initiative* (Washington DC: UNDP) available at http://www.opt-init.org/framework/DOI-Final-Report.pdf.

Dornbusch, R. (1992) "The Case for Trade Liberalization in Developing Countries", *Journal of Economic Perspectives*, 6(1): 69–85.

Dunning, J. H. (1996) "The Geographic Sources of the Competitiveness of Firms: Some Results of a New Survey", *Transnational Corporations*, 5: 1–29.

Duong, T. T. (undated) "Trade and Investment in Vietnam: Towards Deeper Integration into the Region and the World", Paper presented at the International Law Conference in ASEAN Legal Systems and Regional Integration, Asia-Europe Institute Malaysia, www.asia-europe-institute.org.

Dutta, S., L. Bruno and P. Fiona (2003) *The Global Information Technology Report: Readiness for the Networked World*, World Economic Forum (New York: Oxford University Press).

Edquist, C. (1997) *Systems of Innovation: Technologies Institutions and Organizations* (Washington, London: Pinter).

Enos, J. (1982) "The Choice of Techniques v/s the Choice of Beneficiary: What the Third World Chooses", in F. Stewart and J. James (eds), *The Economics of New Technology in Developing Countries* (London: Frances Pinter).

Enos, J. L. and W. H. Park (1988) *The Adoption and Diffusion of Imported Technology: The Case of Korea* (London: Croom Helm).

Ernst, D. (1993) *The Global Race in Micro Electronics* (New York, Frankfurt: Campus).

——(2001) "From Digital Divides to Industrial Upgrading: Information and Communication Technology and Asian Economic Development", Working Paper Economics Series No. 36, East-West center, Honolulu.

——(2002) "Global Production Networks and the Changing Geography of Innovation Systems: Implications for Developing Countries", *Economics of Innovation and New Technology*, 11(6): 497–523.

Ernst, D. and L. Kim (2002) "Global Production Networks, Knowledge Diffusion and Local Capability Formation", *Research Policy*, 31(8–9): 1417–1429.

Ernst, D. and B. A. Lundvall (2000) "Information Technology in the Learning Economy: Challenges for Developing Countries", Working Paper No. 8, East-West center, Honolulu.

Feong, K. W., F. H. Oh and I. Shin (2001) "Economic Impact of Information Communication Technology in Korea", in M. Pohjola (ed.), *Information Technology, Productivity and Economic Growth* (New York: Oxford University Press).

Fransman, M. (1984) "Some Hypotheses Regarding Indigenous Technological Capability and the Case of Machine Production in Hong Kong", in M. Fransman and K. King (eds), *Technological Capability in the Third World* (London: Macmillan).

——(1986) *Technology and Economic Development* (Sussex, Brighton: Wheatsheaf Books).

Fransman, M. and K. King (1984) *Technological Capability in the Third World* (London: Macmillan).

Freeman, C. (1986) *The Economics of Industrial Innovation* (London: Pinter Publishers).

——(1987) *Technology Policy and Economic Performance: Lessons from Japan* (London: Pinter).

——(1995) "The National Innovation Systems in Historical Perspective", *Cambridge Journal of Economics*, 19(1): 5–24.

——(2002) "Continental, National and Sub-national Innovation Systems-Complementarity and Economic Growth", *Research Policy*, 31: 191–211.

Freeman, N. J. (2002) "Foreign Direct Investment in Cambodia, Laos and Vietnam: A Regional Overview", Paper presented in the Conference on Foreign Direct Investment: Opportunities and Challenges for Cambodia, Laos and Vietnam, 16–17 August.

General Statistical Office (2000) *Analyzing the Results of the Industrial Survey of Vietnam – in 1999* (Hanoi: Statistical Publishing House).

——(2001) *Statistical Yearbook: 2000* (Hanoi: Statistical Publishing House).

——(2002) *Statistical Yearbook: 2001* (Hanoi: Statistical Publishing House).

Government of India (2001) *India Knowledge Super Power: Strategy for Transformation* (New Delhi: Planning Commission).

Government of Thailand (2002) *The Ninth National Economic and Social Development Plan (2002–2006)* (Bangkok: Thailand National Economic and Social Development Board, Office of the Prime Minister).

Government of the Union of Myanmar (1998) *The Union of Myanmar: Foreign Investment Law, Procedures and Types of Economic Activities Allowed for Foreign Investment* (Yangon: Government of Myanmar).

——(1999) *Economic Development of Myanmar* (Yangon: Ministry of National Planning and Economic Development).

——(2002) *Selected Monthly Economic Indicators* (Myanmar, Yangon: Central Statistical Organization).

——(undated) *Doing Business in Myanmar* (Yangon, Myanmar: Investment Commission).

——(2001) "Ministry of National Planning and Economic Development", *Statistical Yearbook: 2001* (Myanmar, Yangon: Central Statistical Organization).

Government of Vietnam (2001) *Policies and Measures on the Acceleration of Information Technology Application and Development in Vietnam* (Hanoi: Ministry of Science Technology and Environment).

——(2002) *Policies and Legal Documents on Information and Telecommunication Technology of Vietnam* (Hanoi: The National Steering Committee on Information Technology).

Gu, S. (1999) *Implications of National Innovation Systems for Developing Countries: Managing Change and Complexity in Economic Development* (Maastricht: UNU-INTECH).

Guha, A. and A. S. Ray (2000) "Multinational versus Expatriate FDI: A Comparative Analysis of the Chinese and Indian Experiences", Working Paper No. 58, Indian Council for Research on International Economic Relations, New Delhi.

Gulati, A. (1998) "Indian Agriculture in an Open Economy: Will it Prosper?", in I. J. Ahluwalia and I. M. D. Little (eds), *India"s Economic Reforms and Development: Essays for Manmohan Singh* (New Delhi: Oxford University Press).

Ha, T. (2001) "Vietnam Internet: Challenges and Prospects", *Vietnam Economic News*, 14, May 31, Hanoi.

Hargittai, E. (1999) "Weaving the Western Web: Explaining Differences in Internet Connectivity Among OECD Countries", *Telecommunications Policy*, 23(10–11): 701–718.

Heeks, R. (1996) *India"s Software Industry: State Policy, Liberalization and Industrial Development* (New Delhi: Thousand Oaks; London: Sage Publications).

Hobday, M. (1994) "Export-led Technology Development in the Four Dragons: The Case of Electronics", *Development and Change*, 25(2): 333–361.

——(1995a) "East Asian Latecomer Firms: Learning the Technology of Electronics", *World Development*, 23(7): 1171–1193.

——(1995b) *Innovation in East Asia: The Challenge to Japan* (Aldershot, United Kingdom: Edward Elgar).

——(2002) "Innovation and Stages of Development: Questioning the Lessons from East and South East Asia", Paper prepared for SOM/TEG – Conference at the University of Groningen, The Netherlands: Empirical Implications of Technology-based Growth Theories, August.

IMF (2001) *World Economic Outlook: The Information Technology Revolution*, Washington DC, October.

India, Department of Electronics (1972) *Annual Report* (New Delhi: Department of Electronics).

——(1985) *New Computer Policy* (New Delhi: Department of Electronics).

——(1986) *Policy on Computer Software Exports, Software Development and Training* (New Delhi: Department of Electronics).

India, Department of Telecommunications (2004) *Annual Report* (New Delhi: Department of Telecommunications).

India, Electronics Commission (1971) Proceedings of the National Conference on Electronics Bombay 24–28 March (Bombay: Electronics Commission).

India, Investment Centre (2002) *SIA News Letter: Annual Issue* (New Delhi: India Investment Centre).

India, Ministry of Information Technology (1998) *IT Action Plan* (3 volumes) (New Delhi: National Taskforce on Information Technology and Software Development), available at http://www.ittaskforce.nic.in/.

——(2000) *Action Taken Report of the National Task Force on Information Technology and Software Development, IT Action Plan: Part I* (New Delhi: Ministry of Information Technology).

——(2004) *Guide to Electronics Industry in India* (New Delhi: Ministry of Information Technology).

India, Planning Commission (2001) *India Knowledge Super Power: Strategy for Transformation* (New Delhi: Planning Commission).

Indjikian, R. and D. S. Siegel (2005) "Impact of Investment in IT on Economic Performance: Implications for Developing Countries", *World Development*, 33(5): 681–700.

Institute of Economics (2002) *The Nominal and Effective Rates of Protection in Vietnam: A Tariff-Based Assessment* (Hanoi: Institute of Economics).

Intarakumnerd, P. and P. Panthawi (2003) "Science and Technology Development Towards a Knowledge-Based Economy", in M. M. Makishima and M. Suksiriserekul (eds), *Human Resource Development Towards a Knowledge-Based Economy: The Case of Thailand* (Japan, Chiba: Institute of Developing Economies, Japan External Trade Organization).

Intarakumnerd, P., P. Chairatana and T. Tangchitpiboon (2002) "National Innovation System in Less Successful Developing Countries: The Case of Thailand", *Research Policy*, 31: 1445–1457.

ITU (2002a) *Competition Policy in Telecommunications: The Case of India* (Geneva: International Telecommunication Union).

——(2002b) *Bits and Bahts: Thailand Internet Case Study* (Geneva: International Telecommunication Union).

——(2002c) *Khmer Internet: Cambodia Case Study* (Geneva: International Telecommunication Union).

——(2002d) *Vietnam Internet Case Study* (Geneva: International Telecommunication Union).

——(2003) *Internet on the Mekong: Laos Case Study* (Geneva: International Communication Union).

Jensen, M. B., B. Johnson, E. Lorenz and B. A. Lundvall (2005) "Forms of Knowledge, Modes of Innovation and Innovation Systems", Department of Business Studies (Mimeo: Aalborg University).

Jitsuchon, S. (1994) *Sources of Thailand"s Economic Growth: A Fifty-Year Perspective (1950–2000)* (Bangkok: Thailand Development Research Institute).

Jitsuchon, S. and Sussangkarn (1993) ""Thailand"", in L. Taylor (ed.), *The Rocky Road to Reform: Adjustment, Income Distribution and Growth in Developing World* (Cambridge, MA: MIT Press).

Johnstone, B. (1989) "Taiwan Holds its Lead, Local Makers Move into New Systems", *Far Eastern Economic Review*, 145: 50–51.

Jorgenson, D. W. and K. J. Stiroh (1995) "Computers and Growth", *Economics of Innovation and New Technology*, 3: 295–316.

Joseph, K. J. (1997) *Industry Under Economic Liberalization: The Case of Indian Electronics* (New Delhi: Thousand Oaks; London: Sage Publications).

——(2000) "Foreign Firms and Export Performance: New Empirical Evidence from Indian Industry", *Development Review*, 1(1): 3–23.

——(2002) "Growth of ICT and ICT for Development: Realities of the Myths of Indian Experience", Discussion Paper No. 2002/78, UNU/WIDER, Helsinki.

——(2004a) "State, FDI and Export of Software and Services from India", A Background paper for the World Investment Report: 2004 (New Delhi: Research and Information System for Developing Countries).

——(2004b) "The Electronics Industry", in S. Gokrn, A. Sen and R. R. Vaidya (eds) *The Structure of Indian Industry* (New Delhi and New York: Oxford University Press).

——(2004c) Development of Enabling Policies for Trade and Investment in the IT Sector of the Greater Mekong Sub-region, available at http://www.unescap.org/tid/projects/gms.asp.

——(2004d) "As the Elephant Follows the Dragon: India's Economic Performance under Globalization", Mimeo (New Delhi: Research and Information System for Developing Countries).

——(2005a) "Trade Liberalization and Digital Divide: A South–South Cooperation Perspective", Paper presented at the WIDER Jubilee Conference on "WIDER Thinking Ahead: The Future of Development Economics", Helsinki, 17–18, June 2005.

——(2005b) "Transforming Digital Divide into Digital Dividend: The Role of South–South Cooperation in ICTs", *Cooperation south 102–125*.

——(2005c) "Strategic Approach Towards Promoting International Competitiveness in Knowledge Intensive Industries: Case of Indian Electronics", RIS Discussion Paper No. 88, Research and Information System for Developing Countries, New Delhi.

——(2005d) "Perils of Excessive Export Orientation", in G. Parayil (ed.), *Political Economy of Information Capitalism in India: Digital Divide, Development Divide and Equity* (New York: Palgrave).

Joseph, K. J. and V. Abraham (2005) "Moving Up or Lagging Behind? An Index of Technological Competence in India"s ICT Sector", in A. Saith and M. Vijayabaskar (eds), *ICTs and Indian Economic Development* (New Delhi: Sage Publications).

Joseph, K. J. and K. N. Harilal (2001) "Structure and Growth of India"s IT Exports: Implications of an Export-Oriented Growth Strategy", *Economic and Political Weekly*, 36(34): 3263–3270.

Joseph, K. J. and P. Intarakumnerd (2004) "GPTs and Innovation Systems in Developing Countries: A Comparative Analysis of ICT Experiences in India and Thailand", Paper presented at the Second GLOBELICS Conference, Beijing, October.

Joseph, K. J. and G. Parayil (2005) "India-ASEAN Cooperation in Information Communication Technologies: Issues and Prospects", in N. Kumar, R. Sen and M. Asher (eds), *ASEAN-India Economic Relations: The Road Ahead* (Singapore: Institute of South East Asian Studies).

Joseph, K. J. and M. Sarma (2002) "Slip Between the Cup and the Lips: Analysis of Approved and Realized FDI in India", A Paper presented in the Second National Seminar on Competition, Regulation and Investment: Role in Economic Growth organized by CUTS, Jaipur, 8–9 June, Chennai.

Kalirajan, K. P. and T. Akita (2003) "Income Inequality and Convergence of Income across Indian States", in K. P. Kalirajan and U. Sankar (eds), *Economic*

Reform and the Liberalization of the Indian Economy: Essays in Honour of Richard T. Shand (Cheltenham: Edwards Elgar).

Kannan, K. P. (1997) "Economic Reform, Structural Adjustment and Development in Cambodia", Working Paper No. 3, Cambodia Development Resource Institute, Phnom Penh.

Kapilinsky, R. (2000) "Spreading the Gains from Globalization: What can be Learned from Value Chain Analysis?", IDS Working Paper No. 110, Institute of Development Studies, Sussex.

Kato, Toshiyasu, Kaplan Jeffrey and Sophal Chan (2000) "Enhancing Good Governance In Cambodia", *Cambodia Development Review*, 4(1): 1–3.

Katrak, H. (1985) "Imported Technologies, Enterprise Size and R&D in a Newly Industrializing Country: The Indian Experience", *Oxford Bulletin of Economics and Statistics*, 47(3): 213–229.

Katz, J. (1980) "Domestic Technology Generation in LDCs: A Review of Research Findings", Working Paper No. 35, IDB/ECLA Programme CEPAL Office, Buenos Aires.

——(1984) "Domestic Technological Innovations and Dynamic Comparative Advantage", *Journal of Development Economics*, 16(1–2): 13–37.

Kaushik, P. D. and N. Singh (2004) "Information Technology and Broad-Based Development: Preliminary Lessons from North India", *World Development*, 32(4): 591–607.

Kevin, F. (2002) "Tax Incentives in Cambodia, Laos and Vietnam", Paper presented at the IMF Conference on Foreign Direct Investment: Opportunities and Challenges for Cambodia, Laos and Vietnam, Hanoi, Vietnam, August 16–17.

Kim, L. (1980) "Stages of development in Industrial Technology in a Developing Country: A Model", *Research Policy*, 9: 254–277.

Kline, S. J. and N. Rosenberg (1986) "An Overview of Innovation", in R. Landau and N. Rosenberg (eds), *The Positive Sum Game* (Washington DC: National Academy Press).

Koanantakool, T. H. (2001) *Getting Ready for the New Millennium, What are the Thai Government"s Actions toward the Year 2005?* (Bangkok: NECTEC).

——(2002) "National ICT Policy in Thailand", Paper presented at the Africa Asia Workshop on Promoting Cooperation in Information and Communications Technologies Development, Kuala Lumpur and Penang, Malaysia, March 25–29.

——(undated) *The Internet in Thailand: Our Milestones*, http://www.nectec.or.th/users/htk/milestones.html.

Kokko, A. (1998) "Vietnam-Ready for Doi Moi II?", SSE/EFI Working Paper Series in Economics and Finance No. 286 (Stockholm: Stockholm School of Economics).

Kraemer, K. L. and J. Dedrick (2001) "Information Technology and Economic Development: Results and Policy Implications of Cross-Country Studies", in M. Pohjola (ed.), *Information Technology, Productivity and Economic Growth* (Oxford University Press).

Krishnan, R. T. and G. N. Prabhu (2004) "Software Product Development in India: Lessons from Six Cases", in A. P. D"Costa and E. Sridharan (eds), *India in the Global Software Industry: Innovation, Firm Strategies and Development* (New York: Palgrave Macmillan).

Krueger, O. Ann (1997) "Trade and Economic Development: How We Learn?", *American Economic Review*, 87(1): 1–22.

Kumar, N. (1994) *Multinational Enterprises and Industrial Organization: The Case of India* (New Delhi: Thousand Oaks; London: Sage Publications).

——(1998) "Technology Generation and Technology Transfers in World Economy: Recent Trends and Implications for Developing Countries", in Kumar *et al.* (eds), *Globalization, Foreign Direct Investment and Technology Transfers: Impact on Prospects for Developing Countries* (London and New York: Routledge).

——(2000) "Developing Countries in International Division of Labor in Software and Service Industry: Lessons from Indian Experience", A Background paper for the *World Employment Report 2001* (New Delhi: Research and Information System for Developing Countries).

——(2001a) "Indian Software Industry Development: International and National Perspective", *Economic and Political Weekly*, 36(45): 4278–4290.

——(2001b) "National Innovation System and Indian Software Industry Development", Background paper for World Industrial Development Report, UNIDO.

Kumar, N. and K. J. Joseph (2004) "National Innovation Systems and India"s IT Capability: Are There any Lessons for ASEAN New Comers?", RIS Discussion Paper No. 72/2004, Research and Information System for Developing Countries, New Delhi.

——(2005) "Export of Software and Business Process Outsourcing from Developing Countries: Lessons from India", *Asia Pacific Trade and Investment Review*, 1(1): 91–108.

Kumar, N. and J. Pradhan (2003) "Determinants of Outward Foreign Direct Investment From A Developing Country: The Case of Indian Manufacturing Firms", RIS Discussion Paper No. 44, Research and Information System for Developing Countries, New Delhi.

Kumar, N. and N. S. Siddharthan (1997) *Technology, Market Structure and Internationalization: Issues and Policies for Developing Countries* (London and New York: Routledge).

Kumar, S. (undated) "Foreign Presence in the Field of IT in Vietnam", *Indian Embassy in Vietnam* (mimeo).

Kyaw, Aye. (2002) *IT Needs and Readiness Assessment of the Union of Myanmar* (Myanmar, Yangon: Posts and Telegraphs, Ministry of Communications).

Lal, R. (2002) "Framework and Strategy for ICT for Development–Digital Opportunity Initiative and UNDP Experience", Paper presented in the National Consultation on IT for Development organized by UNDP Vietnam, Melia Hotel Hanoi, December 16.

Lall, S. (1982) "Technological Learning in the Third World", in F. Stewart and J. James (eds), *The Economics of New Technology in Developing Countries* (London: Frances Pinter).

——(1987) *Learning to Industrialize: The Acquisition of Technological Capability in India* (London: Macmillan).

——(1992) "Technological Capabilities and Industrialization", *World Development*, 20(2): 165–186.

——(2001) *Competitiveness, Technology and Skills* (Cheltenham: Edward Elgar).

Lauridsen, L. S. (2002) "Coping with the Triple Challenge of Globalization, Liberalization and Crisis: The Role of Industrial Technology and Technology Institutions in Thailand", *The European Journal of Development Research*, June, 14(1): 101–125.

Link, A. N. and D. S. Siegel (2003) *Technological Change and Economic Performance* (New York, London: Routledge).

List, F. (1841) *The National System of Political Economy*, English Edition (London: Longman, 1904).

Louise, A. and K. Frances (2002) "Higher Education Development", *Cambodia Development Review*, 6(1): 12–17.

Lundvall, B. A. (1985) *Product Innovation and User–Producer Interaction* (Aalborg: Aalborg University Press).

——(1988) "Innovation as an Interactive Process: From User–producer Interaction to the National Innovation Systems", in G. Dosi, C. Freeman, R. R. Nelson, G. Silverberg and L. Soete (eds), *Technology and Economic Theory* (London: Pinter Publishers).

——(ed.) (1992) *National Systems of Innovation: Towards a Theory of Innovation and Interactive Learning* (London: Pinter Publishers).

——(2002) "Editorial", *Research Policy*, 31(2): 185–190.

——(2003) "Why the New Economy is a Learning Economy", Economia Politica Industriale: Rassegna trimestrale diretta da Sergio Vacc'a/Vacc'a, Sergio, Milano: FrancoAngeli s.r.l, 117: 173–185.

Lwin, T. (1999) "GMPS Implementation in Myanmar", Presentation made at GMPS Regional Workshop, Seoul, July 5–7, 1999.

Makino, S., C. M. Lau and R. S. Yeh (2002) "Asset Exploitation versus Asset Seeking: Implications for Location Choice of Foreign Direct Investment from Newly Industrialized Economies", *Journal of International Business Studies*, 33: 403–421.

Malerba, F. (2002) "Sectoral Systems of Innovation and Production", *Research Policy*, 31: 247–264.

Malerba, F. (ed.) (2004) Sectoral Systems of Innovation Concepts, Issues and Analyses of Six Major Sectors in Europe (Cambridge: Cambridge University Press).

Mansell, R. (1999) *Global Access to Information Communication Technologies* (Johannesburg and Canada: International Development Center).

Mathews, J. A. (1999) "From National Innovation Systems to National Systems of Economic Learning: The Case of Technology Diffusion Management in East Asia", DRUID Summer Conference on National Innovation Systems, Industrial Dynamics and Innovation Policy, Rebild, Denmark, June 9–12, 1999.

——(2001) "National Systems of Economic Learning: The Case of Technology Diffusion Management in East Asia", *International Journal of Technology Management*, 22: 455–479.

Mephokee, C. (2003) "Thai Labour Market in Transition Toward a Knowledge-Based Economy", in M. Makishima and Somchai Suksiriserkul (eds), *Human Resource Development Toward a Knowledge-Based Economy: The Case of Thailand*, Institute of developing economies (Japan, Chiba: Japan External Trade Organization).

Micevska (2003) "ICTs for Pro-poor Provision for of Public Goods and Services", Paper presented at Regional Conference on Digital GMS, Asian Institute of Technology, Bangkok, February 26–28.

Millar, J. and R. Mansell (1999) *Software Applications and Poverty Reduction: A Review of Experience*, INKSPRU, University of Sussex.

Mohnen, P. (2001) "International R&D Spillovers and Economic Growth", in M. Pohjola (ed.), *Information Technology, Productivity and Economic Growth* (Oxford University Press).

Murray, M. J. (1980) *The Development of Capitalism in Colonial Indo-China, 1870–1940* (Berkeley: University of California Press).

Mytelka, L. K. and J. F. E. Ohiorhenuan (2000) "Knowledge-based Industrial Development and South–South Cooperation", *Cooperation South*, No. 1: 74–82.

Nagaraj, R. (1997) "What has Happened since 1991? Assessment of India"s Economic Reforms", *Economic and Political Weekly*, 32(44–45): 2869–2879.

Nagaraj, R. (2003) "Industrial Policy and Performance since 1980: Which Way Now?", *Economic and Political Weekly*, 38(35): 3707–3715.

Narayanamurthy, N. R. (2000) "Making India a Significant IT Player in this Millennium", in Romila Thapar (ed.), *India: Another Millennium* (New Delhi: Viking and Penguin Books).

NASSCOM (1999) *Indian IT Software and Services Directory* (New Delhi: National Association of Software and Service Companies).

——*The IT Software and Services Industry in India: Strategic Review 2000* (New Delhi: National Association of Software and Service Companies).

——*Indian IT Enabled Service Providers Directory 2002* (New Delhi: National Association of Software and Service Companies).

——*Indian IT Software and Services Directory 2003* (New Delhi: National Association of Software and Service Companies).

——*The IT Software and Services Industry in India: Strategic Review 2004* (New Delhi: National Association of Software and Service Companies).

Nath, P. and A. Hazra (2002) "Configuration of Indian Software Industry", *Economic and Political Weekly*, 37(8): 737–743.

National Information Technology Committee (1996) IT (2000) "Thailand National IT policy" (Bangkok: Ministry of Science Technology and Environment, National Electronics and Computer Technology Center).

National Institute of Statistics (2002) *Cambodia Statistical Yearbook 2001* (Phnom Penh).

National Institute of Statistics (2002) *Report of the 2001 Manufacturing Industry Survey: Whole Kingdom* (Bangkok: National Statistical Centre).

National Statistical Center (2000) *Basic Statistics about the Socio-Economic Development in the Lao P.D.R.* (Vientiane: National Statistical Center).

National Statistical Center (2002) *Basic Statistics about the Socio-Economic Development in the Lao P.D.R.* (Vientiane: National Statistical Center).

NECTC (2002) *ICT for Poverty Reduction: Examples of Programmes/Projects in Thailand* (Bangkok: NECTEC).

Nelson, R. R. (1981) "Research on Productivity and Productivity Differentials, Dead Ends and New Departures", *Journal of Economic Literature*, 19: 1029–1064.

——(ed.) (1993) *National Innovation Systems: A Comparative Analysis* (Oxford: Oxford University Press).

——(1995) "Recent Evolutionary Theorizing about Economic Change", *Journal of Economic Literature*, 33: 48–90.

Nelson, R. R. and S. G. Winter (1977) "In Search of a Useful Theory of Innovation", *Research Policy*, 16(1): 36–76.

——(1982) *An Evolutionary Theory of Economic Change* (Cambridge, MA: The Belking Press of Harvard University Press).

Neosi, J., P. Saviotti, B. Bellon and M. Crow (1993) "National Systems of Innovation: In Search of a Workable Concept", *Technology in Society*, 15: 207–227.

Nongphanga, C. (1990) "National Information Policy and the Progress of National Information System in Thailand", in Ian Dickson and Lisa Dewyer (eds), *National Information Policies for the Asia and Oceania Region*, 80–81 (Clayton: FID/CAO Secretariat).

Norris, P. (2001) *Digital Divide: Civic Engagement, Information Poverty and the Internet Worldwide* (Cambridge University Press).

NTITSD (1998) *IT Action Plan* (3 volumes), National Taskforce on Information Technology and Software Development, available at http://ittaskforce.nic.in/.

OECD (2000) *Information Technology Outlook* (Paris: OECD).

Oliner, S. D. and D. E. Sichel (1994) "Computers and Output Growth Revisited: How Big is the Puzzle?", Brookings Papers on Economic Activity; *Macroeconomics*, 2: 273–331.

Oranger, S. (2001) "National Information Policy: Differing Approaches", Paper presented in the IT Awareness Seminar organized by NIDA on September 11–13, Phnom Penh.

Page Jr, J. M. (1979) "Small Enterprises in African Development: A Survey", World Bank Staff Working Paper No. 363, The World Bank, Washington DC.

Palasri, S., S. Huter and Z. Wenzel (1999) *The History of the Internet in Thailand* (Eugene: The Network Startup Resource Centre), http://www.nsrc.org/case-studies/thailand/english/index.html.

Parayil, G. (2003) "Digital Divide and Increasing Returns: Contradictions of Informational Capitalism", paper presented in the Seminar on Indian Development organized jointly by the School of Public Policy at George Mason University and the Department of Management Studies, Indian Institute of Science, March 3–5, Bangalore.

——(2005) "Introduction: Information Capitalism", in Parayil, G. (ed.), *Political Economy and Information Capitalism in India: Digital Divide, Development Divide and Equity* (New York: Palgrave).

Parthasarathi, A. and K. J. Joseph (2002) "Limits to Innovation with Strong Export Orientation: The Experience of India"s Information Communication Technology Sector", *Science Technology and Society*, 7(1): 13–49.

——(2004) "Innovation under Export Orientation", in A. P. D"Costa and E. Sridharan (eds), *India in the Global Software Industry: Innovation, Firm Strategies and Development* (New York: Palgrave Macmillan).

Patel, S. J. (ed.) (1995) *Technological Transformation of the Third World*, Vol. 1 (Honkong, Aldershot: Aveburry).

Patibandla, M. D. Kapur and B. Petersen (2000) "Import Substitution with Free Trade: Case of India"s Software Industry", *Economic and Political Weekly*, 35(15): 1263–1271.

Pearce, R. D. (1999) "The Evolution of Technology in Multinational Enterprises: The Role of Creative Subsidiaries", *International Business Review*, 8: 125–148.

Phissamay, P. (2002) *Current Status of IT Development in Laos*, http://goanna.cs.rmit.edu.au/~aym/rinseap/bali/PPT-LaoPDR/.

Pohjola, M. (2001) "Information Technology and Economic Growth: A Cross Country Analysis", in M. Pohjola (ed.), *Information Technology, Productivity and Economic Growth* (New York: Oxford University Press).

Pongsrihadulchai, A. (1998) "Application of Information Technology in Agriculture in Thailand", available at http://www.jsai.or.jp/afita/afita-conf/1998/S01.pdf.

Puntasen, A. *et al.* (2001) *The Demand for IT Manpower in Thailand* (Bangkok (in Thai): National Information Technology Committee Secretariat, National Electronics and Computer Technology Centre).

Quah, D. (2001) "The Weightless Economy in Economic Development", in M. Pohjola (ed.), *Information Technology, Productivity and Economic Growth* (New York: Oxford University Press).

Radosevic, S. (1999) *International Technology Transfer and Catch-up in Economic Development* (Cheltenham: Edward Elgar).

Ratanak, H. (2001) "The Open Forum Information Exchange: A Civil Society IT Contribution to the Development of Cambodia", Paper presented at the Information Technology Awareness Seminar organized by NIDA, September, 11–13, Phnom Penh.

Rattakul, Rattananwan (2003) *Bridging the Digital Divide: A Case Study of CATNET Nationwide Internet Kiosks*, in Proceedings: The Regional Conference on Digital GMS, Asian Institute of Technology, Bangkok.

Razavi, M. R. and A. Maleki (2004) "Applying National Innovation Systems Approach in the Context of Industrializing Countries: Methodological Unity and Terminological Diversity in Literature", Paper presented in the Second GLOBELICS Conference, Beijing.

Reserve bank of India (2004) *Handbook of Statistics on Indian Economy*, 2002–2003, (Mumbai: Reserve Bank of India).

RIS (2004) *ASEAN-India Vision 2020: Working Together for a Shared Prosperity* (New Delhi: Research and Information System for Developing Countries).

Rodrik, D. (1992) "The Limits of Trade Policy Reform in Developing Countries", *Journal of Economic Perspectives*, 6(1): 87–105.

——(2004) "Rethinking Growth Policies in the Developing World", Lucad" Agliano Lecture in Development Economics, delivered on October 8, 2004, in Torino, Italy, available at http://courses.umass.edu/econ804/rodrik2.pdf.

Rodrik, D. and A. Subramanian (2004a) "From Hindu Growth to Productivity Surge: The Mystery of the Indian Growth Transition", IMF Working Paper No. WP/04/77, International Monetary Fund, Washington DC.

——(2004b) "Why India can grow at 7 Percent a Year or More: Projections and Reflections", IMF Working Paper No. WP/04/118, International Monetary Fund, Washington DC.

Romer, P. M. (1993) "Ideas Gaps and Objects Gaps in Economic Development", *Journal of Monitory Economics*, 32: 543–73.

Saggi, K. (2002) "Trade Foreign Direct Investment and International Technology Transfer: A Survey", *The World Bank Research Observer*, 17(2): 191–235.

Saxenian, A. (2002) "The Silicon Valley Connection: Transnational Networks and Regional Development in India Taiwan and China", *Science Technology and Society*, 7(1): 117–149.

Sichel, D. E. (1997) *The Computer Revolution: An Economic Perspective* (Washington DC: Brookings Institution Press).

Singh, D. (2002) "Electronics Commerce: Issues of Policy and Strategy for India", ICRIER Working Paper No. 86 Indian Council for Research on International Economic Relations, New Delhi.

Singh, N. (2002) "Crossing a Chasm: Technologies, Institutions and Policies for Developing a Regional IT Industry", Paper presented in the International Seminar on ICTs and Indian Development organized by ISS, The Hague and IHD, Banglore, December 9–11.

Soete, L. (2000) "Towards the Digital Economy: Scenarios for Business", *Telematics and Informatics*, 17, 199–212.

——(2001), "Electronic commerce and the Information Highway", in Louis-André Lefebvre, Elizabeth Lefebvre and Pierre Mohnen (eds), *Doing Business in the Knowledge-based Economy: Facts and Policy Challenges*, Kluwer Academic Publishers, 307–327.

Solow, R. M. (1987) "We'd Better Watch out", *New York Times Book Review*, July 12.

Sreekumar, T. T. (2002) "Civil Society and ICT-Based Models of Rural Change: History, Rhetoric and Practice", Paper presented at the International Seminar on "ICTS for Indian Development: Processes, Prognosis and Policies" jointly organized by Institute of Social Studies, 9–11 December, The Netherlands and Institute of Human Development, New Delhi, Bangalore.

Srinivasan, T. N. and J. N. Bhagwati (1999) "Outward Orientation and Development: Are Revisionists Right?", Discussion Paper No. 806, Economic Growth Center, Yale University, New Haven.

Steinmueller, W. E. (2004) "The European Software Sectoral System if Innovation", in Malerba, F. (ed.), Sectoral Systems of Innovation Concepts, Issues and Analyses of Six Major Sectors in Europe (Cambridge: Cambridge University Press).

Stiglitz, J. E. (2002) *Globalization and Its Discontents* (New York: Allen Lane).

Subrahmanian, K. K. (1995) "Technological Transformation: An Assessment of India's Experience", in S. J. Patel (ed.), *Technological Transformation of the Third World*, Vol. 1 (Honkong, Aldershot: Aveburry).

Suksiriserekul, S. (2003) "ICT Manpower Development", in M. Makishima and Suksiriserkul, Somchai (eds), *Human Resource Development Toward a Knowledge-Based Economy: The Case of Thailand* (Japan, Chiba: Institute of Developing Economies, Japan External Trade Organization).

Sukumar, S. and N. R. Narayanamurthy (2003) "Influence of ICT on the Development of India's Competitiveness", in Dutta, S. *et al.* (ed.), *The Global Information Technology Report: Readiness for the Networked World* (New York: Oxford University Press).

Tangkitvanich, S. (2001) "State Intervention in the Internet Market: Lessons from Thailand", Paper presented at the ITU, Workshop on Internet in South East Asia, November 21–23, Bangkok, Thailand.

Tangkitvanich, S. and T. Ratananarumitsorn (2002) "Competition and Regulatory Reform in the Thai Telecommunication Markets", *TDRI Quarterly Review*, 17(4): 112–118.

TDRI (1992) *The Development of Thailand's Technological Capability in Industry* (Bangkok: Thailand Development Research Institute).

Thajchayapong, P., H. Reinermann and S. E. Goodman (1997) "Social Equity and Prosperity: Thailand Information Technology Policy into the 21st Century", *The Information Society*, 13: 265–286, available at http://www.fes.uwaterloo.ca/crs/plan674d/informtnsoc1312k.pdf.

Thampi, S. (2003) "ICT Development Trends and Needs for its Further Growth in GMS", Key Note address delivered at Regional Conference on Digital GMS, February 26–28, Asian Institute of Technology, Bangkok.

Thompson, V. (1937) *French Indo China* (London: Macmillan).

Thoraxy, H. (2003) *Cambodia"s Investment Potential: Challenges and Prospects* (Japan: International Cooperation Agency, Phnom Penh).

Tipton, F. B. (2002) "Bridging the Digital Divide in Southeast Asia: Pilot Agencies and Policy Implementation in Thailand, Malaysia, Vietnam and the Philippines", *ASEAN Economic Bulletin*, 19(1): 83–99.

TRAI (2005) *Study Paper on Indicators for Telecom Growth* (New Delhi: Telecom Regulatory Authority of India), http://www.trai.gov.in/ir30june.pdf.

TuyHoa (2001) "Vietnam Internet: Challenges and Prospects", *Vietnam Economic News*, No. 14, May 31, Hanoi.

UNCTAD (1995) *Incentives and Foreign Direct Investment* (Geneva: Background Report of UNCTAD Secretariat).

——(different years) *World Investment Report* (Geneva, New York: United Nations).

——(2000) *World Investment Report: Crossborder Mergers and Acquisitions and Development* (New York: United Nations).

——(2003) *E-Commerce and Development Report 2003* (Geneva: United Nations).

——(2004a) *Pre-Conference Negotiating Text* (Geneva: United Nations).

——(2004b) *E-Commerce and Development Report 2003* (Geneva: United Nations).

UNDP (1999) *Human Development Report 1999* (New York: UNDP/Oxford University Press).

——(2002) *E-Readiness Assessment in the Laos* (Laos: UNDP/UNV Vientiane).

UNESCAP (1998) *Enhancement of Trade and Investment Cooperation in South East Asia: Opportunities and Challenges Toward ASEAN-10 and beyond* (New York: United Nations).

——(1999) *Alignment of the Trade Documents of Cambodia Myanmar and Viet Nam*, Studies in Trade and Development No. 38, United Nations, New York.

——(2000) *Private Sector Perspectives in the Greater Mekong Subregion* (New York: United Nations).

——(2001) *Greater Mekong Subregion Business Handbook* (Bangkok: United Nations).

——(2002) *Trade Facilitation Handbook for the Greater Mekong Subregion* (New York: United Nations), p. 3.

——(2004a) *Economic and Social Survey of Asia and the Pacific 2004: Asia Pacific Economies Sustaining Growth and Tackling Poverty* (New York: United Nations).

——(2004b) Trade and Investment Policies for the Development of the Information and Communication Technology Sector of the Greater Mekong Sub region, Studies in Trade and Investment No. 52 (New york: United Nations).

UN ICT Taskforce (2005) *Innovation and Investment: Information Communication Technologies and the Millennium Development Goals* (New York: UN ICT Taskforce).

UN Task Force on Science, Technology and Innovation (2005) *Innovation: Applying Knowledge in Development* (London and Sterling Va: Earthscan).

US Department of Commerce (2000) *Digital Economy 2000 Report* (Washington, DC: US Department of Commerce).

——(2002a) *Country Commercial Guide: Burma*, http://www.usatrade.gov/website/ccg.nsf/ShowCCG?OpenForm&Country=BURMA.

——(2002b) *Country Commercial Guide: Laos*, http://usembassy.state.gov/Laos/wwwhcotg.html.

——(2002c) *Country Commercial Guide for Vietnam*, http://www.usatrade.gov/Website/ CCG.nsf/ShowCCG?OpenForm&Country=VIETNAM.

——(2002d) *Country Commercial Guide for Cambodia*, http://www.usatrade.gov/Website/CCG.nsf/ShowCCG?OpenForm&Country=CAMBODIA.

——(2004) *Country Commercial Guide*, http://www.usatrade.gov/website/ccg.nsf/ShowCCG?OpenForm&Country=THAILAND.

USAID (2001) *Vietnam: ICT Assessment Final Report*, USAID Asia Near East (ANE) Bureau Sponsored Study, September.

Vietnam Competitiveness Initiative (2003) Software/ICT Cluster strategy, J. E. Austin Associates Vietnam Representative Office, Hanoi.

Viotti, E. B. (2001) "National Learning Systems: A New Approach on technical Change in Late Industrializing Economies and Evidences from the Cases of Brazil and South Korea", *Science, Technology and Innovation*, Discussion Paper No. 12, Center for International Development, Harvard University, Cambridge, MA, USA.

Wade, R. (1990) *Governing the market: Economic Theory and the Role of Government in East Asian Industrialization* (Princeton, NJ: Princeton University Press).

Walsh, K. (2003) *Foreign High-Tech R&D in China: Risks, Rewards, and Implications for US-China Relations* (Washington, DC: The Henry L. Stimson Center).

Weidman John, C. (1997) "Laos, in Gerard A. Postiglione and Grace", C. L. Mak (ed.), *Asian Higher Education: An International Handbook and Reference Guide* (CT, Westport, USA: Greenwood Press).

Wignaraja, K. (2002) "ICT as an enabler for Development", Presentation made at the National Consultations on ICT for Development Organized by UNDP Vietnam, December 16, Hanoi.

Wilson, D. and R. Purushothaman (2003) "Dreaming with BRICs: The Path to 2050", Global Economics Paper No. 99 (New York: Goldman Sachs).

Wong, P. K. (2001) "The Contribution of Information Technology to the Rapid Economic Growth of Singapore", in M. Pohjola (ed.), *Information Technology, Productivity and Economic Growth* (New York: Oxford University Press).

Wong, P. K. and A. A. Singh (2002) "ICT Industry Development and Diffusion in Southeast Asia", in C.S. Yue and J. J. Lim (eds), *Information Technology in Asia: New Development Paradigms* (Singapore: Institute for South East Asian Studies).

World Bank (1999) "Knowledge for Development", *World Development Report 1998–99* (New York: Oxford University Press).

——(2000) *Korea and the Knowledge Based Economy: Making a Transition* (OECD: World Bank).

——(2002) *Information Communication Technology: A World Bank Group Strategy* (Washington DC: The World Bank).

——(2003) "Vietnam: Deepening Reforms for Rapid Export Growth", Synthesis Report for Vietnam"s Export Study, World Bank, Washington DC.

——(2004) *World Development Indicators* (Washington DC: World Bank).

Xaphakdy Smith, *Developing Information and Communication Infrastructure: An Implementation of Tokyo Declaration and Action plan in Laos*, available at http://unpan1.un.org/intradoc/groups/public/documents/APCITY/UNPAN006169.pdf.

Yue, C. S. and J. J. Lim (eds) (2002) *Information Technology in Asia: New Development Paradigms* (Singapore: Institute of South East Asian Studies).

Index

Action Plan, 172–5
  for innovation, 209
  for IT development, 175
  for software development, 174
adaptive R&D, 15, 17
  see also incremental improvements
ADB, 149, 161, 169, 170, 231
ADSL, 80
Advance Micro Devices, 229
Advanced Information System, 78
advances in ICT and "splintering
  off", 8
affordability of ICT, 10
  innovations for, 16
AFTA, 114, 128, 129, 189
agro-based industries competitiveness
  of, 130
Alcorta, L. and Peres W., 14
All India Council of Technical
  Education, 41
Alphatec Group, 229
AMD, 54
animation multimedia potential of, 58
Annual survey of Industries, 35
Arnold, E. et al, 59, 65
Arocena R. and Sutz, J., 13, 14
Arora, A., 25, 38
Arora, A. and Athreya S., 29
Arora, A. and Gambardella A., 16
Arora, A. et al, 3
ASEAN, 4, 53, 54, 128, 159, 226
  development of electronics in, 57
  entry by Myanmar, 141
  entry by Vietnam in, 170
  joining of Laos in, 114
  lessons for, 205
ASEAN help ASEAN strategy, 218
ASEAN New Comers, 4, 208
  foreign direct investment in, 187
  fructification of FDI in, 71
  lessons to be learned by, 208

performance of Vietnam compared
  to other, 201, 202
promoting IT use and neglecting IT
  production, 214
reforms undertaken by, 211
Asian Institute of Technology,
  Bangkok, 232
assembly-oriented industries, 10
Association of Freight Forwarders, 81
Asynchronous Transfer Mode, 80
Australian National University, 235
automatic approval, 212
automatic route, 48
  see also automatic approval

backward and forward Linkages, 56
Bagan Cybertech, 166
Bagan Cybertech and FTP service, 150
Bagan Cybertech and MPT mail
  users, 149
Bago River from Yangon, 149
balance of payments, 100
Bali Summit, 4
Bangladesh, 8
basic telecommunication, 16
Basu, K., 227
BCC, 235
Bell, M. and Pavitt, K., 14
Bell, M. et al, 14
Bezanson K. J. et al, 171, 172, 234
Bhabha Committee, 39
Bhagwati, J. N., 226
Bhide, S. and Kalirajan K. P., 227
Bhoomi Land Records, 135, 215
Bigpond, 108, 109
Bilateral Trade Agreement, 170
Bimal Sareen, 232
Binh and Houghton, 187
Board of Investment, 67, 71
Bora, B., 223
bottom-up approach, 121, 136
Breschi, S. and Malerba, F., 13

Bresnahan, T. and
    Trajtenberg, M., 226
Bretton Woods Institutions, 86
Brooker group, 59
BTA, 187, 234
    increase in FDI due to, 189
budget deficit growing, 141
Build-Transfer Contract, 184
Build-Transfer-Operate Scheme, 78, 79,
    184, 230
Business Cooperation Contracts, 182
Business India, 228
Business Process Outsourcing, 8
    characteristics of, 25
    growth in India, 23
    opportunities for developing
        countries in, 219
    source of employment and export, 3
Business Support Centre, 42
business to business, 49

Cairns and Nikomborirak, 75
Cal Comp, 54
Cam GSM, 91
Cambodia Mobile Telephone
    Company, 92
Cambodia Samart Communications
    Company Ltd, 92
Cambodia Shinawatra, 90
CamGSM, 107
Camintel, 92, 104
Camnet, 108
Camshin, 91, 104, 108
    major player in mobile market
        Cambodia, 92
Camtel, 91, 107
Canadian International Development
    Agency, 170
Cannon, 54
Capability Maturity Model, 24, 62
capacity building, 20
capacity utilization, 15
capital goods, 55
capital intensive, 6
CAPNET, 236
Carlos, A. P. *et al*, 226, 232
Carlsson, B. S. *et al*, 13
CasaCom, 106, 107
Cascom, 91

CAT, 76, 81, 229
    monopoly of, 79
    role in SchoolNet, 82
    see also School Net.
catching up, 14
    by China, 17
    with ICT revolution, 4
CATNET, 232
CDC, 96, 230
Cellular Mobile Services, 43
Center for International Economics,
    100, 190, 235
Central Executive Committee of the
    Communist Party of Vietnam, 174
Central Statistical Office, 148
Central Statistical Organization,
    35, 232
Centre for Development of Advanced
    Computing, 44
Chamber of Commerce, 81
Charoen Pokphand Group, 229
Chief executive Officers Program, 80
Chief Information Officer, 81
Chile, 8
China, 21
    Chunhui Program, 192
    emerging provider of IT and BPO,
        8, 16
Chinese in Silicon Valley, 191
Chu Huu Quy, 169
CIO program, 80
CISCO, 94
Civil Servant Commission, 81
Civil Society Organizations, 111
    addressing digital divide, 210
    pooling resources of, 213
clearance of machinery imports, 69
Code Division Multiple Access, 149
codified knowledge, 16
commercial production, 50
Committee for Planning and
    Cooperation, 132
commodity problematique, 7
Common Effective Preferential Tariff
    Scheme, 159
Communication Authority of
    Thailand, 232
    monopoly of, 78
company-based training centers, 65

comparative advantage, 10
  of local capital, 126
Completely Knocked Down, 129
Computer Science Development
  Council, 142
  duties and powers of, 232
Computer Science Development
  Law, 142
Computer Society of India, 228
concentrated market structure, 7
concessionaries, 230
connectivity, 16
Cooper, C., 14
CorDECT Wireless in Local Loop, 16
  addressing connectivity, 215
  lowest cost connectivity
    solution, 236
Costa Rica, 16
Council for the Development of
  Cambodia, 95
Council of Scientific and Industrial
  Research, 44
CP group of Thailand, 92
cross-border certification, 16
cross-border movement of labor, 7
crowding-in effect, 126
CSDC IT master plan of
  Myanmar, 145
  organization of, 233
CSO, 158, 162, 232, 233
current account deficit, 21

Dahlman, C. J., 14
Dahlman, C. J. and Brimble, P., 59
Dahlman, C. J. and Westphal, L., 14, 15
Daknet, 135
  asynchronous broadband
    linkage, 215
Danang (DSP), 179
David, P. A., 1
D'Costa, A. P., 3, 30
DDFI, 123
DeBression, C. and Amesse, F., 13
DeBresson, C., 13
Decree of the Prime Minister, 122
  attracting investment in IT, 127
Department of Computer Science
  and Automation, 44

Department of Domestic and Foreign
  Investment, 122
Department of Electronics, 38, 40
  accreditation Society, 41
  accreditation system, 67
  formation of, 39
  promotion of R&D, 44
  setting up of STP, 42
Department of Posts and
  Telecommunications, 116
Department of Skill
  Development, 65
Department of
  Telecommunications, 43
Desai, A. V., 15
design-intensive goods, 7
destabilized economy, 86
Dewang Metha, 228
diffusion, 5
  of ICT in India, 30
  of ICT in manufacturing, 34
  of mobile technology, 163
digital divide, 5
  Among ASEAN New
    Comers, 219
  capacity building to bridge, 217
  cooperation to reduce, 216
  limits of ITA to address, 19
  see also international digital divide
digital dividend, 19
Digital Gangetic Plain, 135
Digital GMS, 232
Digital Opportunity Initiative, 2, 8
digital threat to development, 2
direct benefits of ICT, 5, 6
direct employment, 8
diversified systems, 13
Doi Moi, 184, 201
domestic availability of technology, 6
domestic R&D effort, 15
Dornbusch, R., 9
Dunning, J. H., 12
Duong, T. T., 235
Dutta, S. *et al*, 4, 37, 224

e-ASEAN Framework Agreement, 5, 20,
  215, 216, 217, 219
e-ASIA, 215
e-choupal project, 35

e-commerce, 81, 146, 147
  developing new markets
    through, 131
  growing use of, 58
  in IT Master Plan, 85
  Permission for FDI in, 49
  in private sector, 83
e-governance, 3, 35, 208, 217, 227, 233
  assistance of India to Laos in, 120
  improving administration
    through, 63
  in IT Master Plan, 85
  ministry of STEA's role in, 115
  role of MPTC in, 152
  role of public private partnership in
    promoting, 146
e-society, 85, 217
e-South Framework Agreement, 19
  for bridging digital divide, 219
  for harnessing southern innovation
    system, 206
East Asian countries, 21
East Asian financial crisis, 20
economic performance, 21
  of Cambodia, 86
  of India, 20
  of Laos, 113
  of Myanmar, 139
  of Thailand, 53
  of Vietnam, 169
Economic Times, 227, 228
EDI, 82
EDPs, 137, 212
Edquist, C., 13
educational management information
    systems, 94
EHTPs, 51
e-learning, 166, 179
Electronic Commerce Resource
    Center, 81
Electronics and Software Export
    Promotion Council, 228
Electronics Commission, 39, 227
Electronics Hardware Technology
    Park, 48
electronics industries in Southeast
    Asia, 56
Electronics Research and Development
    Centre, 44

empowering people, 8
Enos, J., 14, 15
Enos, J. L. and Park, W. H., 14
Enterprise of Post and
    Telecommunications Lao, 117
Enterprise of Telecommunications
    Lao, 117
Entrepreneurial Development
    Program, 127
Entrepreneurship, 9
Entrepreneurship Development
    Centre, 120
Entrepreneurship Development
    Institute of India, 81
entry barriers, 6
Ernst, D., 6, 7, 11, 16, 17, 56
Ernst, D. and Kim, L., 10
Ernst, D. and Lundvall, B. A., 11
ETL, 116
European Union, 189
evolutionary approach, 13
exchange of information, 5
  for capacity building, 219
export enclaves, 30
export of ICT, 5
export of technology, 17
export-orientation, 25
export-oriented production, 11
Export Oriented Units, 48
Export Promotion Department of
    Thailand, 129

Faculty of Engineering and
    Architecture, 119
FDI, 12, 103, 125, 153
  in and IT and manufacturing, 187
  changing approach towards, 67
  incentives in Vietnam for, 178
  inflows in Laos, 124
  inflows to Cambodia, 97
  inflows to India, 43, 50
  inflows to Myanmar, 156
  inflows to Thailand, 70
  inflows to Vietnam, 186
  opening up for, 128
  policies in Cambodia, 89
  policies in India, 38, 47, 48, 49
  policies in Laos, 121
  policies in Myanmar, 152

FDI – *continued*
  policies in Vietnam, 184
  preferred mode of, 167
  promotion in IT Laos, 114
  role in Myanmar economy, 157
  role in Thailand, 59
  sector wise distribution
    Cambodia, 98
  sector wise distribution of, 71
  see also MNEs; transnational
    corporations (TNCs)
FEA, 231
Federation of Thai Industry, 81
Feong, K. W. *et al*, 3
FIL, 122
  incentives offered by, 123
  objectives of, 153
  promotion of private
    investment, 152
File Transfer-Protocol, 150
FIPB, 49
fixed line telecommunication tariff in
    Thailand, 77
foreign and domestic investment in
    Cambodia, 97
Foreign Exchange Certificates, 160
foreign exchange neutrality, 48
Foreign Exchange Regulation Act, 47
foreign exchange reserves, 21
Foreign Investment Management
    Committee, 123
Foreign Investment Promotion
    Board, 48
foreign-owned enterprises, 122
foreign technology collaboration
    agreement, 49
France, 22
Fransman, M., 14, 15
Free/Open Source Software, 110,
    135, 215
Freeman, C., 12, 13, 15, 123, 184, 185,
    187, 235
Frequency Allocation Act, 79
Fujitsu, 54, 228

Gambia, 8
Gazette of India, 41
GDP, 21

structure and growth in
    Cambodia, 86
structure and growth in India, 22
structure and growth in Laos, 113
structure and growth in
    Myanmar, 139
structure and growth in
    Thailand, 53
structure and growth in
    Vietnam, 169
GE Capital, 17
General Department of Posts and
    Telecommunications, 182
General Purpose Technology, 1, 5,
    16, 226
General Statistical Office, 186, 188,
    193, 194, 199, 234
General System of Preferences, 97
geographical dispersion, 7
GERD, 59
global brand building and plans, 27
global integration, 1
global knowledge, 9
global production networks, 7, 10,
    11, 17
global trading environment, 4
globalization, 7
GLOBELICS, 226
Globenet, 134
Globenet set-up, 118
GMS Forum, 231
Golden Jubilee Network, 79
Government Information
    Network, 80
Government of Cambodia, 92
granting permission for
    foreigners, 69
grey market, 165
growth breeds inequality, 131
growth-led export, 210
growth performance of ICT sector, 18
GSM, 92
GSM technology, 133
GSM-based operations, 195
GSP, 99
Gu, S., 13
Gulati, A., 227
Gyan Doot programme, 8

Hanoi People's Committee, 179
hardware innovations, 16
Hargittai, E., 236
HCL, 27
Heeks, R., 26
hidden costs, 96
high-speed communication lines, 80
high value-adding skill-intensive
    activities, 207
Ho Chi Minh City, 179, 180, 181,
    195, 234
Hobday, M., 12, 56
horizontally disintegrated market
    segments, 7
human capital, 7, 12, 16, 20
human resource development
    Program, 61

IBM, 54
ICT, 1, 204, 226
    access to, 2
    as a GPT, 5
    for catching up, 16
    challenges in Thai sector, 55
    for development, 12, 35
    diffusion in India, 34, 37
    diffusion of, 10, 224
    export boom in India, 30
    global trade in, 222
    goods, 6, 16
    induced growth and human
        welfare, 2
    induced productivity growth, 8
    industries, 6
    infrastructure in Cambodia, 89
    infrastructure in Myanmar, 147–51
    infrastructure in Vietnam, 181–3
    investment in Lao PDR, 127
    investment in Thailand, 71
    investment in Vietnam, 187
    investments in, 1
    policy in Cambodia, 88
    policy in India, 38–44
    policy in Lao PDR, 114
    policy in Myanmar, 141–7
    policy in Thailand, 60–7
    policy in Vietnam, 170–80
    power house, 51
    production in Cambodia, 103

    production in India, 23
    production in Myanmar, 165
    production in Thailand, 54
    production in Vietnam, 190–3
    production of, 3, 5
    returns to, 8
    returns to capital in, 3
    returns to investments in, 8
    returns to production of goods in, 6
    returns to production of services
        in, 7
    returns to use of, 8
    revolution, 6
    services, 6
    strategy for developing
        countries, 16
    use in Cambodia, 104–9
    use in India, 33
    use in Lao PDR, 115–18, 131–5
    use in Myanmar, 161–4
    use in Thailand, 75–82
    use in Vietnam, 194–201
    use of, 4, 5, 213
ICT-induced development, 6
IDRC, 170
IDRC-CIDA, 178
I-Flex, 27
IMF, 8
imitation, 15
Imperial Tobacco Company, 35
import capacity, 7
import-substituting growth
    strategy, 86
improving social service provision, 8
IMR, 25
incentive competition, 125
income and employment
    opportunities, 7
income-earning opportunities, 205
incremental improvements, 15
Index of Claimed Technological
    Competence, 27
India Infotech Centre, 42
India–ASEAN Partnership, 4, 5
India–ASEAN Vision 2020, 4
Indian Institute of Science, 44
Indian Institute of Technology, 44
Indian software enterprises, 24, 26, 27
indirect benefits of ICT, 6

Indjikian, R. and Siegel, D. S., 2, 226
inducement mechanisms and focusing
    devices, 14
industrial upgrading, 11
information exchange, 1
Information Markup Language, 232
Information Technology Agreement,
    3, 4, 18, 51, 221, 224, 226, 237
  effectiveness of, 222
  members vs non-members of, 223
  trade liberalization under, 225
  see also WTO
Information Technology Center with
    Value Addition, 120
Infosys, 24, 27
initial project analysis, 68
Initiative for ASEAN Integration, 5
initiatives for integration, 216
    limits to ASEAN help ASEAN, 218; *see
      also* ASEAN help ASEAN strategy
    transition to ODM and OBM, 217;
      see also ODM; Own Brand
      Manufacture (OBM)
innovation system, 9, 12, 202, 205, 209
  and capability building, 16
  investment location and, 71
  trade and investment policies and,
    18, 54
  see also National System of
    Innovation
innovative process, 15
instability in commodity prices, 7
Institute of Economics, 189
institutional and infrastructure
    support, 13
institutional and policy
    environment, 60
institutional infrastructure, 21
institutional interventions, 12
institutional linkages, 26
Intarakumnerd, P. *et al*, 13, 58
Intarakumnerd, P. and
    Panthawi, P., 61
integrated circuits, 54
Intel, 6
Inter-University Network, 82
interlinked system of research
    centers, 9

International Committee for
    Reconstruction of Cambodia, 86
International Development Research
    Center, 115
international digital divide, 77, 214
international division of labor, 8, 10
international Export Promotion
    Organizations, 129
international production network, 12
International Technical Exchange
    Cooperation Program, 120
International Trade Center, 129, 223
Internet Connecting Providers, 175
Internet Service Providers, 27, 49, 92,
    134, 232
  foreign ownership in, 79
  in private sector, 119
internet services by LaoTel, 118
intra-ASEAN trade in ICT
    products, 217
intra-national digital divide, 2, 34, 51,
    214, 224
  civil society to mitigate, 64, 210
  in developing countries, 77
  internet use in urban areas leading
    to, 136, 213
intra-national telecom divide, 132,
    162, 214
  inter-regional variations in
    teledensity and, 195
  reduction in, 201
investing in Pentium or in
    Penicillin, 208
investment-led growth, 218
investment liberalization, 204
Investment Promotion Act, 67
Investment Promotion Law, 121
IPR, 47
iPSTAR, 91
ISDN, 80
ISO 9000, 68
IT business associations, 180
IT enterprise development, 89
IT infrastructure, 89
IT Master Plan, 60, 208
  in Myanmar, 145
  by STEA, Lao PDR, 118
  for Thailand, 66
IT security, 16

Italian-Thai Development, 229
IT-enabled service, 24, 25, 26,
    103, 212
  in Lao PDR, 137
  manpower for, 42
  opportunities of income and
    employment from, 219
  outsourcing of, 206
ITU, 2, 43, 75, 92, 108, 134, 181, 194,
    224, 232, 235

Japan, 12, 16
Japan External Trade
    Organization, 129
Jasmine and Loxley, 229
Java and NET, 229
Jeffrey A. Bader, 234
Jensen, M. B. *et al*, 15, 16
Jhai foundation, 215, 231
John A. Dally, 232
John C. Weidman, 231
Johnstone, B., 57
joint ventures, 71, 154, 157, 158,
    167, 184
  crowding in of foreign investment
    due to, 126
  entry barriers for investment in, 137
  foreign investments in, 119, 122, 211
  investment by, 75
  to facilitate international trade, 82
Jorgenson, D. W. and Stiroh, K. J., 1
Joseph, K. J., 2, 16, 29, 31–7, 46, 204,
    226- 228
Joseph, K. J. and Abraham, V., 3, 27
Joseph, K. J. and Harilal, K. N., 31
Joseph, K. J. and
    Intarakumnerd, P., 62
Joseph, K. J. and Parayil, G., 5

Kannan, K. P., 95, 230
Kaplinski, R., 7
Karnataka, 9
Kato, T. *et al*, 87, 89
Katrak, H., 15
Katz, G., 14, 15
Kaushik, P. D. and Singh, N., 9, 227
Khmer language in computers, 208
Kim, L., 14, 15
know-how, 14

knowledge-based asset-seeking
    strategies, 12
knowledge-based economy, 5, 63
knowledge-intensive industries,
    8, 38
knowledge-intensive sectors, 38
Koanantakol, T. H., 16, 79
Kokko, A., 169, 234
Kraemer, K. L. and Dedrick, J., 5, 6
Krishnan, R. T. and Prabhu, G., 227
Krueger, A. O., 9
Kumar, N., 15, 24, 25, 27, 42, 44
Kumar N. and Joseph K. J., 17, 27, 30
Kumar, N. and Pradhan, J. P., 22
Kumar, N. and Siddharthan, N. S., 14
Kyats, 140
Kyaw, Aye, 151, 163, 164

Lall, S., 11, 14, 16
Lao-Japan Technical Training
    Center, 119
Lao National Internet Committee,
    119, 232
Lao-Shinawatra Telecom
    Company, 117
Lao Telecom Asia Co. Ltd, 116
LaoTel, 134
last mile connectivity, 133
Lauridsen, L. S., 58, 65, 229
learning by doing, 15
Lee Thorn, 232
liberalized policies, 26
liberalized trade policy regime, 12, 55
limits under WTO, 19, 216
  exports of good under, 219
Link A. N. and Siegel, D. S., 2
Linux-based graphical desktop, 215
List, F., 12
Local Area Network, 151, 201
local capabilities, 11
local content, 16
Local Loop Technology, 215
look East policy, 4
Low Skill-intensive and low
    value-adding activities, 25
LTC, 116
Luang Namtha, 115
Lucent, 54

Lucent Technologies
    Microelectronics, 229
Lundvall, B. A., 12, 13

McKinsey, 22
macroeconomic benefits of IT
    production, 6
macroeconomic stabilization, 86
Madhya Pradesh, 9
Malerba, F., 13
man-hour charges, 27
Mangolia, 24
Mansell, R., 3
Manufacturers Association of
    Information Technology, 228
market mediated technology
    transfer, 15
market-oriented economic
    management, 184
market-oriented liberalization, 100
market-oriented reforms, 139
market segments, 7
Mastek, 24, 25
Mathews, J. A., 13
MCA program, 120
Media Lab Asia program, 135
medical transcription, 110, 168
Memorandum of
    Understanding, 120
Mephokee, C., 56
Micevska, 116
Micro Info Centre, 119
microprocessors, 6
microwave connections and digital
    exchanges, 149
Milicom Lao Co.Ltd, 116
Military Electronic
    Telecommunications
    Company, 182
Millicom, 91, 117
Ministry of Commerce, Cambodia, 96
Ministry of Commerce,
    Myanmar, 159
Ministry of Communication,
    Myanmar, 144
Ministry of Communication, Post and
    Construction, Lao PDR, 116
Ministry of Information
    Technology, 209

Ministry of Information Technology,
    India, 41, 42, 44, 49, 50
Ministry of Labor and Social Welfare,
    Lao PDR, 123
Ministry of Post and Telematics,
    Vietnam, 182
Ministry of Science, Technology and
    Environment, Vietnam, 170
Multinational Corporations (MNCs),
    12, 17, 67, 71, 125, 165
    complimentary capabilities for,
        167, 211
    expansion of operation of, 58
    in stages of development, 57
    see also MNEs; transnational
        corporations (TNCs)
MNEs, 17, 227
    budget on R&D, 21
    diaspora effect on, 25
    exports, 30
    see also Multinational Corporations
mobile communication
    technology, 87
Mobitel, 108
MoEYS, 94
Monopolies and Restrictive Trade
    Practice Act, 47
monopoly in international
    services, 78
monopoly of Caminco, 102
Morocco, 16
Most Favored Nation, 97, 99,
    221, 234
MOU, 179
MPT, 162, 163
MPTC, 90, 92, 104, 110, 141, 152
Multi Fibre Agreement, 97
multi-hop Wi-Fi network, 135
Mutual Recognition
    Arrangements, 217
Myanmar Chamber of Commerce and
    Industry, 159, 165
Myanmar Computer Enthusiasts
    Association, 143
Myanmar Computer Entrepreneurs
    Association, 143
Myanmar Computer Industry
    Association, 165

Myanmar Computer Science Development Council, 141
Myanmar Foreign Investment Law, 152
Myanmar ICT Development Corporation, 166
Myanmar Investment Commission, 152, 154
Myanmar Machine Tool and Electrical Industries, 165
Myanmar Post and Telecommunications, 145, 161, 162, 233
Mytelka, L. K. and Ohiorhenuan, J. F. E., 204

NAFTA, 234
Nagaraj, R., 227
Narayana Murthy, N. R., 40
NASSCOM, 23, 24, 28, 47, 228
Nath, P. and Hazra, A., 34
National Assembly of the Socialist Republic of Vietnam, 182
National Bank of Cambodia, 96
National Broadcasting Commission, 78–9
National Center for Natural Science and Technology, 171
National Centre for Software Technology, 44
National Chamber of Commerce and Industry, 129, 130
National Commission on Electronics, Bombay, 227
National Conference of Electronics, 39
National Economic and Social Development Plan, 53
National Education Network Project, 83
National Electronics and Computer Technology Center, 60
National Information Infrastructure, 91, 60
National Information Technology Committee, 60, 63, 81, 82
National Information Technology Development Authority, 87
national innovative infrastructure, 46

National Institute of Statistics, 95, 97
National Long Distance Service, 43
National Science and Technology Development Agency, 62
National Statistical Center, 125, 230
National System of Innovation, 12–14, 38, 58, 84, 227
  in an evolutionary perspective, 13, 206
  for harnessing new technology for development, 37
  in the IT sector, 39, 51, 64, 170
  strengthening of, 63, 66, 130, 141, 207
  see also regional systems of innovation; sectoral systems of innovation
National Task force, 35
National Task Force on Information Technology and Software Development, 41
National Telecom Policy, 43
National Telecommunication Commission, 78
National University of Lao PDR, 119
National University of Laos, 119
NECTEC, 8, 80, 81, 82, 230
Nelson, R. R., 13
Nelson, R. R. and Winter, S. G., 13, 15
network of institutions, 12
Network Readiness Index, 4, 37
New Economic Mechanism, 114, 128
new industrial policy, 51
Newly Industrializing Countries, 14, 16
NGOs, 3, 86, 93, 108, 200, 203, 207, 231
  capacity building and IT training by, 94
  harnessing ICT for rural masses, 64
  harnessing technology for development, 204
  public–private partnership, 89
NIDA, 94, 95, 209
NIIT, 24, 179
Non ICT capital, 8
non-market mediated technology transfer, 15
Non-Resident Indians, 51, 206, 224
Nongphanga Chitrakorn, 229

normal trade relations, 234
Norris, P., 236
North American Free Trade
    Agreement, 189
North West region, 195
North–South cooperation, 220
not only born but also
    developed, 212
NRI/OCB, 48
NS Electronics, 54, 229
NTT West, 229
NYNEX, 229

ODA to modernize the telecom
    facilities, 117
ODM, 12, 56, 58, 217
OECD, 8
offshore development, 26
Oliner, S. D. and Sichel, D.E., 1
Open Forum of Cambodia, 93
open-market economy, 128
Operational Service Providers, 175
opportunity cost, 38
Oranger, S., 88
Original Equipment Manufactures,
    12, 57
output elasticity of ICT capital, 8
outsourcing of services, 6, 8
outward-oriented growth, 86
Overseas Corporate Bodies, 48
Own Brand Manufacture (OBM),
    12, 58, 217

Pacific and South Asia, 224
Page Jr, 15
Parayil, G., 2, 204
Parthasarathi, A. and Joseph K. J.,
    38, 42, 191, 226
Patel, S. J., 14, 15
patent, 21
Patibandla, M. *et al*, 52
patterns of innovation, 14
pent-up demand, 2
pharmaceuticals, 21
Phissamay, P., 134, 232
Planetonline, 118, 134
Planning Commission of India, 9
Pohjola, M., 2, 8
Polaris, 24

policy of renovation (Doi Moi), 169
Pongsrihadulchai, A., 230
poverty alleviation, 7–8
Power Lobbying, 228
primary producing countries, 7
pro market reforms, 20
process industries, 10
product design engineering, 14
production planning engineering, 14
productivity paradox, 226
Promotion and Management of
    Foreign Investment, 122
promotion certificate issuance, 69
proprietary software products, 27
PT Indosat of Indonesia, 90
Public Internet Exchange, 79
PubNet, 79
PVK Computer Center, 120

radical innovations, 15, 17
    see also Schumpeterian type of
        innovations
Radio Paging, 43
Radosevic, S., 14
Ratanak, H., 93
Rattakul Rattananwan, 232
Rattana Business Administration
    College, 120
Razavi, M. R. and Maleki, A., 13
RBI, 48, 49
Red River delta, 195
regional dispersion, 7
regional governments, 26
regional systems of innovation, 13
    see also National System of
        Innovation; sectoral systems of
        innovation
remittance and FDI, 161
Research and Development (R&D),
    21, 57
    centers, 17
    expenditure, 59
    intensity, 27
    investment in, 33
    outsourcing, 17
    policy in electronics, 39
Research and Technology
    Organization, 62
RGC, 95

RIS, 5
RMIT, 234
Rodriguez, 236
Rodrik, D., 11, 20
Rodrik D. and Subramanian A., 21, 22
role of multilateral
    organizations, 127
Royal Government of Thailand, 64
Royal Group of Cambodia, 91
Royal Melbourne Institute of
    Technology, 179
rule-based multilateral trading
    System, 159
rural ICT projects, 9

Saggi K., 12
Samart Group of Thailand, 92
Santacruz Electronics Export
    Processing Zone, 39
SATCOM, 233
satellite telephones, 163
Satyam Computer Services, 27
Saxenian, A., 192
scale down technology, 15
scale intensive, 6, 10
School Net., 82
Schumpeterian type of
    innovations, 13
    see also radical innovations
Science Technology and Environment
    Agency, 114
Seagate, 54
sectoral dialogue partnership, 4
sectoral systems of innovation, 13
    see also National System of
        Innovation
segmentation of the industry, 6
Semi Knocked Down, 129
Service Provider Agreement, 163
service sectors, 17
SGS to conduct Pre-shipment
    Inspection on goods, 102
Shin Corporation, 90
Shin Satellite PLC, 90
Sichel, D. E., 1
Siem Reap, 108
Silicon Valley, 42, 227, 228
Silverline, 24
Simputer, 16, 215

Simputer Trust, 136
Singapore, 8
Singapore Telecommunications Ltd., 43
single window clearance system, 111,
    138, 147, 213
skill-intensive, 22
skill requirements, 6
Skill upgradation and knowledge
    generation, 84
skilled manpower, 21, 22
SMEs, 27, 123, 154, 180
socio-economic transformation, 9
software, 21
    and BPO, 21
    piracy, 47
    promotion scheme, 39
    value chain, 3
    ventures, 25
Software Engineering Institute, 24
Software Park of Thailand, 61, 62, 84
Software Technology Parks, 26, 42,
    150, 166, 178, 179, 202, 206, 209
    export oriented units in, 51
    foreign investment in, 48
    objectives of, 228
    policy for, 52
Solow, R. M., 226
South Asia, 147, 148, 162, 224
South Asian Countries, 181
South East Asia, 11, 16, 134, 163,
    169, 230
    penetration of internet in, 109
    telephone density in, 131
Southern capabilities, 5
Southern innovation system, 14–18,
    214, 215
    for catching up, 205, 206
    effective use of, 19
    untapped potential of, 219
Southern problems, 5
South–South cooperation, 205, 215,
    219, 220
    see also Southern innovation system;
        e-South Framework Agreement
Special Economic Zones, 48, 50, 51
specialized skills, 6, 11
spillover benefits, 5
splintering off, 8

Srinivasan, T. N. and
    Bhagwati, J. N., 9
stabilization-cum-structural
    adjustments, 20
Stankiewiez, 13
state economic enterprises, Myanmar
    Economic Holdings Ltd, 157
State Law and Order Restoration
    Council, 233
state-owned Economic
    Enterprises, 159
state-owned enterprises, 114
STEA, ministry of, 115, 118, 208
STI mode of innovation, 15
sticky specialization, 11
Stiglitz, J. E., 11
STPs, 235
strategy of walking on two legs, 210
structural retrogression, 139
structural transformation, 21, 53
Sub-Saharan Africa, 224
Subrahmanian K. K., 14, 15
Sudhir Kumar, 234
Suksiriserekul, S., 229
Sukumar S. and
    Narayanamurthy. N. R., 236
Summit-level interaction, 4
Super-computer Education and
    Research Centre, 44
sustained growth, 11
system for information/IT managers
    (CIO), 176

Tangkitvanich, S. and
    Ratananarumitsorn, T., 75, 77, 78
TARAhaat and Drishtee, 227
tariff and non-tariff barriers, 9
Tata consultancy services, 27, 35
Tata Institute of Fundamental
    Research, 44
technological capability, 7, 14, 15
technological intensity, 6, 16, 17
technological systems, 13
technology-generating effort, 59
technology licensing, 15
technology park scheme, 50
TEDIC, 81, 82
Telecom Asia, 76, 229
telecom density, 105, 106

telecom sector, 26
telecommunication expansion
    program, 60
Telecommunication Regulatory
    Authority, 61
telephone density, 2
Telephone Organization
    Thailand, 78
Texas Instruments, 28
Thai Farmers Bank, 229
Thai Industrial and Innovation
    survey, 65
Thai Ministry of Education, 65
Thai Social/Scientific, Academic and
    Research Network, 79
Thailand EDI Council, 81
Thajchayapong, 229
Thampi, S., 114
Thanlyin Satellite Communication
    Station, 149
Thoraxy, H., 99
Time Division Multiple Access, 150
Tipton, F. B., 77
top-to-bottom approach, 121
top-to-down approach in
    education, 136
Toshiba and Daewoo, 165
TOT, 76, 80, 81, 82, 229
trade, 10
    liberalization, 20
    performance in Lao PDR, 130
    policies in Lao PDR, 128
    policies in Myanmar, 158–60
    policies and performance in
        Cambodia, 99
    policies in India, 50–1
    policies in Thailand, 69
    policies in Vietnam, 189
trade and investment, 4, 9, 12, 18
    liberalized regime in, 11
    limits to liberalization in, 11
    policies and performance in
        Cambodia, 95–102
    policies and performance in
        Thailand, 67–71
    policies in India, 47–51
    policies in Lao PDR, 121–31
    policies in Myanmar, 152–60
    policies in Vietnam, 184–90

TRAI, 33, 43
Training the Trainers, 120, 136, 210
transaction cost, 1, 5
transfer of technology, 15
transnational corporations (TNCs), 11, 218
see also MNEs; Multinational Corporations (MNCs)
transport-intensive goods, 131
TRI, 92
TT&T, 76
Tuy Ha, 191

US-based ICT players, 6
UN ICT Taskforce 2005, 2, 224
UN Task Force on Science, Technology and Innovation, 2
UNCTAD, 2, 7, 16, 70, 97, 125–30, 156, 185, 186, 190, 204
UNDP, 115, 117, 118, 119, 128, 132, 133, 231
UNESCAP, 53, 83, 109, 117, 118, 231, 235
UNESCO, 94
United Nations Transnational Authority in Cambodia, 90
Urban–rural Digital Link, 115
US Department of Commerce, 6, 56, 58, 96, 124, 231, 235
US PTO, 17
USAID, 179, 180, 200

value-added approach, 230
Value Added Services, 43
value chain, 7, 12
VAT, 103
Venture Capital, 25, 42
vertically integrated, 7
Very Small Aperture Terminals, 78
based network, 120
network, 150
services, 43
Vientiane Municipality and Vientiane Province, 113
Vientiane summit, 4
Vietnam Association of Information Processing (VAIP), 180
Vietnam Chamber of Commerce and Industry, 180

Vietnam Post and Telecommunications Corporation, 181
Vietnamese Labor Supply Organization, 185
Vietnam's Post and Telecommunication development, 183
Vinaphone and Mobifone, 195
Viotti, E. B., 13
VNPT, 182, 234, 235
Vocational Training promoting act, 65
volatile global finance, 11
VSNL, 43

Walsh, K., 17
WHO, 108
Wi-Fi, 215
Wide Area Network, 176, 201
Wilson, D. and Purushothaman, R., 22
WIPRO, 27
wireless fidelity, 135
Wireless in Local Loop, 92
Wong, P. K., 3, 8
Wong, P. K. and Singh, A. A., 226
wood-based industries, 130
World Bank, 1, 8, 9, 139, 147, 148, 149, 162, 164, 190
World Summit on Information Society, 204
WTO, 3, 114, 141
conditional compliance, 128
entry into, 170, 189
and ITA, 51, 216, 218, 219, 221
limits to ITA, 18, 19
observer status in, 129
tariff protection commitments, 69
telecom liberalization as per, 78
see also Information Technology Agreement

Xaphakdy, Smith, 231

Y2K problem, 207
Yangon City Development Committee, 157
Yue C. S. and Lim, J. J., 3